# ROUTLEDGE LIBRARY EDITIONS: THE ECONOMY OF THE MIDDLE EAST

Volume 7

# ARAB OIL POLICIES IN THE 1970s

# ARAB OIL POLICIES IN THE 1970s
Opportunity and Responsibility

YUSIF A. SAYIGH

LONDON AND NEW YORK

First published in 1983

This edition first published in 2015
by Routledge
2 Park Square, Milton Park, Abingdon, Oxon, OX14 4RN

and by Routledge
711 Third Avenue, New York, NY 10017

*Routledge is an imprint of the Taylor & Francis Group, an informa business*

© 1983 Yusif Sayigh

All rights reserved. No part of this book may be reprinted or reproduced or utilised in any form or by any electronic, mechanical, or other means, now known or hereafter invented, including photocopying and recording, or in any information storage or retrieval system, without permission in writing from the publishers.

*Trademark notice*: Product or corporate names may be trademarks or registered trademarks, and are used only for identification and explanation without intent to infringe.

*British Library Cataloguing in Publication Data*
A catalogue record for this book is available from the British Library

ISBN: 978-1-138-78710-0 (Set)
eISBN: 978-1-315-74408-7 (Set)
ISBN: 978-1-138-80996-3 (Volume 7)
eISBN: 978-1-315-74501-5 (Volume 7)
Pb ISBN: 978-1-138-82007-4 (Volume 7)

**Publisher's Note**
The publisher has gone to great lengths to ensure the quality of this reprint but points out that some imperfections in the original copies may be apparent.

**Disclaimer**
The publisher has made every effort to trace copyright holders and would welcome correspondence from those they have been unable to trace.

*YUSIF A. SAYIGH*

# ARAB OIL POLICIES IN THE 1970s

**Opportunity and Responsibility**

CROOM HELM
LONDON & CANBERRA

©1983 Yusif Sayigh
CROOM HELM LTD
Provident House, Burrell Row, Beckenham, Kent BR3 1AT

British Library Cataloguing in Publication Data

Sayigh, Yusif A.
 Arab oil policies
 1. Petroleum industry and trade — Arab countries
 I. Title
 338.2'7'28209174927    HD9578.A55

ISBN 0-7099-2374-0

Printed and bound in Great Britain

# CONTENTS

| | |
|---|---|
| List of Tables | vii |
| Preface | ix |
| Acknowledgement | xiii |
| **1 Identifying Major Policy Areas** | 1 |
| **2 The Policy of Control** | **19** |
| Backdrop to the 1970s | 19 |
| Why Policy Control? | 24 |
| Why Not Earlier Policy Control? | 32 |
| How Was Control Achieved? | 45 |
| Instrumentalities of Control | 55 |
| **3 New Policy Options in an Integrated Context** | **62** |
| Upstream Operations: Policies and Implications | 65 |
|    Exploration | 65 |
|    Production/Conservation | 77 |
|    Marketing | 94 |
|    Pricing | 101 |
| Downstream Operations: Policies and Implications | 139 |
|    Refining | 142 |
|    Gas Treatment | 147 |
|    Petrochemical Industries | 156 |
|    The Economics of Downstream Operations | 169 |
|    The Infrastructure of Downstream Operations | 172 |
|    The Politics of Downstream Operations | 173 |

| | |
|---|---|
| **4 Oil as Engine of Development** | 178 |
| The Financial Implications of the Policy of Control | 178 |
| The Developmental Impact of the Policy of Control | 198 |
| **5 Opportunity and Responsibility** | 219 |
| Introduction | 219 |
| The National and Regional Context | 220 |
| The International Context | 233 |
| Notes | 248 |
| Index | 267 |

## TABLES

3.1: Proven Crude Oil Reserves in Arab Oil-exporting Countries and World, 1979 and 1980 — 67
3.2: Comparison of Exploration Activity in Various Regions of the World during the 1970s — 74
3.3: Arab Oil-exporting Countries' Crude Oil Production, 1979-80 — 80
3.4: Capacity of Refineries in Existence, Under Construction, and in Planning in Arab Oil-exporting Countries — 144
3.5: Natural Gas Reserves of the Arab Oil-exporting Countries, End 1979; Gas Production, Utilisation, and Flaring, 1979 — 148
3.6: Capacity of Gas Liquefaction Plants in Existence and Planned, 1980 — 152
3.7: World Reserves of Fossil Fuels — 154
3.8: Investments in Oil and Gas-related Projects in the Arab Oil-exporting Countries — 159
3.9: Organic Petrochemical Products, in Existence, Under Construction, or in the Planning Phase in the Oil-exporting Arab Countries, 1980 — 161
3.10: Capacity of Existing Plants and Plants Under Construction or in the Planning Stage for the Production of Ammonia in Oil-exporting Countries, 1980 — 162
4.1: Official Prices of One Barrel of Arabian Light Crude — 185
4.2: Oil Production and Revenues of the Seven Arab Oil-exporting Countries, 1961-79 — 181
4.3: Resources and Resource Uses of Seven Arab Oil-exporting Countries, 1979 — 180
4.4: Sectoral Distribution of Planned Investments, 1970-80 — 194

*PREFACE*

I presume to believe that, until mid-October 1973, few people in the advanced, industrial countries, or in the developing countries of the Third World, stopped to think of the issues and complexities involved in the control, production, management, pricing, and marketing of the oil resource. To them, oil was gasoline to feed cars with, fuel-oil to operate central-heating systems, or kerosene for the primuses to cook on in the poorer households. Then, suddenly, with steep increases in oil prices, the oil industry became a matter of general discussion, in most of its aspects, and the questions of volume of production of oil, its price, and its availability in the markets jumped to the forefront of people's preoccupations and discussions everywhere. And as suddenly, OPEC (the Organisation of Petroleum Exporting Countries), of which probably no more than one adult out of every hundred thousand had heard before that decisive October day, was a front-page topic in every leading newspaper in the world and a prominent subject of analysis and comment in the programmes of radio and television stations at least in the advanced industrial countries.

Unfortunately, the speed with which the media took up the subject of oil pricing and production policies, and that of OPEC, was far greater than the speed with which real understanding of these policies, and of the Organisation which formulated them, was acquired. No doubt a sincere and sustained effort was undertaken by a number of newsmen, economic analysts, and politicians to acquaint themselves with the facts and intricacies of the subject of the political-economy of oil. But, judging by much of the

*Preface*

commentary that has flowed in print and on the air since the autumn of 1973, neither the number of those seeking better understanding, nor the depth of the understanding acquired, have proved adequate to clear much of the confusion and misunderstanding that have engulfed the issues of the supply and price of oil.

Furthermore, the Arab oil exporters and their oil policies have received much criticism, and have been the central target of a massive campaign of condemnation, although they are by no means the sole owners and suppliers of oil.

It would not be far fetched to say that the singling out of the Arab exporters for criticism has been in fact the product of a complex of factors — cultural, historical, political, and economic — which combined and interacted, and finally led to the oversimplification that all the blame for the rise in oil prices was to be heaped on the Arab producers. It would be rare to find a Western oil consumer, even today., who realises that his government levies a tax on a barrel of oil imported and refined by his own country, larger than the producing government charges for the barrel of crude exported. It would be equally rare to find a consumer who realises that the question of the price and supply of oil is but one aspect — and not the major aspect, at that — of the much broader and many-sided global question of total energy with which the whole world has to contend: the availability of the various known sources of energy, the development of new sources, the finiteness and non-renewability of fossil energy, and the huge financial implications of the effort to assure humanity of energy for its future.

In brief, the Arab oil policies of the 1970s, the decade which witnessed the major changes in oil policies, need much better and wider understanding, if the confusion and condemnation surrounding these policies is to be cleared. That it is essential to achieve a much larger measure of mutual understanding between oil exporters (among whom the Arabs are important) and oil importers is beyond doubt in my mind. Deep misunderstanding can only place both parties on a course of confrontation and conflict. I submit that a course of co-operation and equitable interdependence is much more in the interest of both parties — both economically and politically. Hence this book, which is hoped to serve as a modest contribution to the improved understanding without which co-operation and interdependence cannot be established.

As the reader will find out, the book is not a wholesale, undiscriminating endorsement of Arab oil policies in the 1970s. Indeed, it contains some sharp criticism of certain aspects of these policies. I wonder, in the end, if it will be the Arab oil-exporting countries or the large oil-importing countries that, on balance, will be more displeased with the book. But if it can clarify the policies it examines, and gain better appreciation by its readers of the reasoning and realities behind them, it would amply justify the work put into it.

The last chapter of the book sets out to establish that the massive, wide-ranging opportunities which the new oil policies have opened up for the Arab countries, in the national, the regional, and the international context, invoke concomitant responsibilities of great magnitude, in the same contexts — responsibilities that can be discharged only if the opportunities are allowed proper scope for fruition. The examination and clarification of this dialectical relationship is hoped to improve further understanding between oil importers and exporters.

One final word is called for. I have tried my utmost to be objective and to write a balanced book. However, I must state from the start that I am an Arab and not a Martian. Therefore, I start from a position of basic sympathy with much of the reasoning and many of the considerations that have shaped Arab policies in the decisive decade of the 1970s. But it is, I hope, responsible, not undiscerning sympathy. Consequently, I run the risk of dissatisfying the critic of Arab policies who refuses to see any merits in them, as well as the supporter who refuses to see any fault in them.

*Preface*

May this book, then, be a useful bridge between the unbending critic and the uncritical supporter of Arab oil policies.

Beirut                                                                Yusif A. Sayigh
April 1982

## ACKNOWLEDGEMENT

I would like to express my thanks to the Petroleum Information Committee of the Arab Gulf States for the research grant they gave me which enabled me to prepare this book and for the complete freedom I enjoyed in writing it. However, I alone bear the responsibility of all the views expressed and the judgements made.

Y.A.S.

# 1

## *IDENTIFYING MAJOR POLICY AREAS*

The 1970s can readily be accorded the distinction of being the most significant decade in the history of Arab oil,* which is more than half a century old. At its most obvious and immediate, this significance is evidenced in the record level of crude oil production attained and prices charged. But it can be evidenced more meaningfully in the exercise of the power of policy formulation and implementation which the Arab oil-exporting countries** has finally appropriated as one of the rightful aspects and instruments of sovereignty over their resources. This power is of particular importance owing to the wide-ranging value of the oil resource in modern life, the substantial reserves and production of Arab oilfields, and the far-reaching implications of control over oil on the national, regional, and international planes.***

It is in this latter context of control that the Arab oil exporters have, at last, registered their most substantial achievement. This is to transform their status from that of

---

*Unless otherwise indicated, the term 'oil' as used in this book refers broadly to hydrocarbon resources in general, that is, oil and gas.

**The term 'oil-exporting Arab countries' refers to the seven Arab members of OPEC, the Organisation of Petroleum Exporting Countries (namely, Algeria, Iraq, Kuwait, Libya, Qatar, Saudi Arabia, and United Arab Emirates). On the other hand, the term 'oil-producing countries' includes the former seven as well as Bahrain, Egypt, Oman, Syria, and Tunisia. All except Oman and Tunisia are members of OAPEC, the Organisation of Arab Petroleum Exporting Countries. Reference to the oil countries in this book will mean the large producers and exporters which are OPEC members, unless otherwise specified. (Since the completion of this book, Tunisia has joined OAPEC, making the membership 11 instead of 10.)

***Throughout, we refer to the 21 Arab states as the components of the Arab region. (The adjective 'regional' is the equivalent of the Arabic term *qawmi*, or pertaining to the Arab homeland in its entirety, whereas the adjective 'national' is the equivalent of *qutri*, pertaining to one *qutr* or individual country.)

the largely passive or powerless receivers of oil policies formulated and implemented outside the purview of their national sovereignty, to that of the sole authority to formulate and implement policies relating to the wide spectrum of issues and activities that fall within the oil industry broadly defined. The shift in the locus of power that this transformation means carries with it an array of changes of considerable weight, both of direct and indirect nature, which it will be our task to identify and assess later on. But at this point, it is essential to indicate that the radical shift in the locus of power and control has meant a commensurately large reallocation of the rewards of the oil industry between the giant foreign oil companies, the producers of the old regime, and the oil countries, the new producers. Furthermore, it is contended here that this shift has been characterised by the success of the oil countries in effecting a judicious blend of their national interests with a sense of international responsibility.

The subsequent chapters of the present book will be devoted to the examination of the major policy areas which have witnessed the changes to which we have alluded, the process through which these were brought about, and the assessment of the impact of these changes and their implications for the future of the oil countries, the Arab region, and the oil countries' international relations. But it is necessary at this early stage to identify the policy areas that will occupy the central court of the discussion in the book. It must be pointed out, however, that although the book is not designed to trace the history of the events preceding the 1970s, a proper appreciation of the policies of the 1970s calls for some acknowledgement of the developments of relevance to policy-making in the period behind the watershed of the early 1970s which led to the later developments. This would put the performance of the 1970s into prominence. For it is vital to point out that the mere fact that the significant advances made in the decade of the 1970s are already a part of the contemporary history of the Arab oil industry, should not obscure the other weighty fact that these advances were possible only after a much longer process of frustrated powerlessness; slow, marginal, and painfully-rationed victories; and finally bitter confrontation and struggle. The hard-won independence of the will of the national states in the sphere of oil policy formulation and decision-making had to be extracted directly from the strong and tenacious grip of the major concessionary Western oil companies, and — by projection — indirectly from that of the industrial powers behind them.[1] It is necessary to keep this perspective in mind because what is now viewed in most quarters as an evident and natural right of the Arabs in the area of oil policies, had been hotly challenged and denied down to October 1973, and in certain respects and in some circles continues to be so down to the present.[2]

We use the term 'policy' in this study to mean a course of action mapped, or a coherent set of guidelines laid down by government, to govern the behaviour of a certain sector or department or area of activity. Obviously, there are good and bad policies, in the sense that inherently policies could *ex ante* be considered by some groups as being built on unsound principles or reasoning, or of faulty design, or aimed at some objective believed to be harmful to society. Alternatively, policies could be considered right from the start to have the opposite qualities and to merit approbation. In the second place, policies could be declared good or bad *ex posti*, that is after they have been allowed the time to prove themselves distinctly beneficial or harmful on balance, irrespective of what they had initially been meant (and/or declared) to be. But, generally speaking, foresight and the good of society are implied in the use of the term 'policy', hence the explanation given in one leading dictionary of 'policy' as being 'prudence, foresight, or sagacity in managing or conducting, especially State affairs; political wisdom, sagacity, artifice, or cunning statecraft; prudent conduct', and finally 'a course of action or administration recommended or adopted by a party, Government, etc'.[3]

It should be admitted at this early stage that the use of the term 'policy' in the present book is in general embedded in overall approval, subject to the strength of criticism that may be directed at specific policies in the course of the analysis. In other words, the discussion starts from the position that the assumption by the Arabs in the 1970s of the power to formulate oil policies pointed on the whole in the 'right' direction — that is, in a direction that represented a large measure of convergence between national, regional, and international interests, though the three sets of interests were not necessarily of equal magnitude (nor could they legitimately be expected to be). Furthermore, it is proposed that the policies adopted could be demonstrated to serve substantial long-term economic, social, and energy purposes that are legitimate and warranted.

But to say this is not to assert that Arab oil policies are an unparalleled, exemplary product of sagacity, wisdom, or prudence and foresight, as the dictionary definition just quoted suggests. Again, it ought to be admitted that, as in all cases where governments or communities of governments lay down policy, there is always room for better application of policy, for greater concern with the content of policy, or for the avoidance of some harm to certain groups caused by policy — to name only a few possibilities for the improvement of policy. In the present context, it is contended that Arab oil policies in the 1970s have a *prima facie* case to be seriously considered and examined, whether in terms of their content and reach, or their impact and implications, and that the examination would justify the initial contention. This statement is defensible in spite of the three reservations that we will hasten to register here. In any case, it will be put to the test of analysis subsequently, like the other propositions submitted in the course of this introductory chapter.

The first reservation is definitional. It relates to the legitimacy of the use of the embracing-yet-exclusive term 'Arab policies'. To begin with, the policy categories or areas which will be shortly identified are not exclusively Arab policies. Virtually all of them, on examination, will be found to be non-Arab as well as Arab in sponsorship, inasmuch as they can be traced to the policy repertoires of the members of OPEC who hail from the continents of

Asia, Africa, and Latin America, and are 7 Arab and 6 non-Arab.* And some policies, like those in the areas of pricing and conservation of hydrocarbon resources, are generally shared by non-OPEC producers such as Norway, the United Kingdom, Canada, and Mexico.

A second reservation applies specifically to Arab oil exporters. This is that several of the policies that will be examined under the heading of 'Arab oil policies' have not been conceived, formulated, or managed collectively by the Arab exporters, whether as members of OPEC or of OAPEC, the Organisation of Arab Petroleum Exporting Countries. Instead, they have been conceived, formulated, and managed by individual governments, though in many instances discussed and agreed upon collectively within OPEC.

This point is worth pursuing a little further. It is interesting to note that although formally certain policies may have fallen totally within a national circumference from conception to implementation, yet a large measure of collectivity in sponsorship can still be discerned. This is because the various oil countries form a definite and identifiable community of interests and expectations, and no doubt a 'field of potential rivalry' as well. They are therefore extremely sensitive to each other's frame of mind, outlook and conduct. It is little wonder, in such a climate, to see policy mainstreams flowing where initial sponsorship and formulation were strictly set within the jurisdiction and scope of action of individual governments. We witness here what can possibly be discerned as an area of ambivalence between approbation and jealousy, both leading to emulation, but both resulting in an observable collectivity of policy conception, formulation, and implementation.

The question, How Arab is Arab oil policy? needs to be stressed in the light of the preceding two paragraphs, for the appreciation of the appropriateness of the use of the embracing term 'Arab' in the discussion to follow, where closer examination will reveal no deliberate and conscious *collective*

---

*The non-Arab members are Ecuador, Gabon, Indonesia, Iran, Nigeria, and Venezuela.

consideration of and decision on several of the policies or policy areas embraced by the analysis. Indeed, again as we shall see later, much of what falls under the umbrella designation 'Arab oil policies' will be found to have been conceived and formulated within OPEC, and not OAPEC which is exclusively Arab in composition — if at all collectively decided upon and formulated. (There will be more to say in this regard when the locus of policy-formulation will come under examination in the next chapter.)

The third reservation relates to the degree of consciousness and purposefulness in policy-formulation. Arab oil policies are probably no different from most economic policies anywhere in the world, in the sense that they are not except rarely the outcome of a formal process of overall, integrated reasoning and structuring, which starts with general conceptualisation and the definition of broad objectives, moves on to the setting of priorities, proceeds to the elaboration of action strategies, designs the plans and programmes called for, and finally formulates the policies for the guidance of action in the direction of the objectives initially determined. Although such a rigorous course of action is rarely followed, except perhaps in the process of more sophisticated development planning (and military planning), it is even rarer to encounter it in the case of the oil policies to be examined in the present study. There is overwhelming evidence to indicate that in the vast majority of cases important policies have grown in discrete steps and stages, to reach their 'critical mass' after a long process of approximation and/or trial and error.

Yet this need not detract from the significance of the policies. The absence of explicit, pre-determined objectives and strategies need not, and often does not, mean that there are no implicit objectives and action strategies that can be imputed and assessed. Yet there is here certainly a serious gap — the failure to externalise; its seriousness derives from the vital but widely ignored principle that there should be much wider participation, by the societies of the countries concerned, in the thinking and debate on oil (as on other) policies. In our view this issue is of crucial and central significance for the future of Arab society (both national and regional).

To go back to the frequent failure to define objectives and action strategies clearly before formulating guideline policies: we believe it is fair to warn that one must not lose sight of the fact that the Arab oil-exporting governments have only had roughly one decade — the 1970s — during which to run through their apprenticeship in the exacting art of control, planning, and management of a most important and complex resource. This apprenticeship implies the mastering of the related political, legal, economic, financial, technical, executive, and managerial skills called into application in the many areas and aspects of oil policies.

What, then, are the major policy areas which, the three reservations notwithstanding, fall in the centre of our attention in this book? These areas are defined by us to include oil policies strictly and narrowly defined, as well as ones that derive from them. The distinction, and the justification of this breadth of scope, will become clear as we proceed. This distinction, it will be seen, is based essentially on the distance of the activity or policy area from the oil sector. For instance, the determination of the volume of production and the pricing of the unit of oil exports are policy areas very closely and directly related to the sector, as against development policy which is more distantly and loosely related, inasmuch as in its early phases development is mainly influenced by the oil industry at a second remove, via the revenues made possible to the treasury of an oil-exporting country, thanks to the production and export of oil.

The justification of the inclusion of development policy in the area of our concern is not much less clear or more difficult to accept. Indeed, the illustration of a seemingly more remote policy area, such as that of development, will be seen to be very relevant to a probing and far-reaching examination of oil policies, if there is a conviction, first, that the end uses of the revenues made possible through the production and pricing policies are of significance to the overall assessment of oil policies, and, secondly, that the close integration of the oil sector with the entire economy is a matter of first priority for the country. We propose that such conviction is soundly based and fully warranted, particularly in the case of countries that are still at a low level of economic development and depend heavily on the oil industry, such as those constituting the Arab region. The argument would be weaker if applied to Norway, Canada, or the United Kingdom as oil producers.

With the broad terms of reference thus adopted, eleven major policy areas can be identified and put under examination in the present study. They are listed here, then defined or identified briefly:

1. The policy of control
2. Exploration for oil
3. Production and the determination of the volume of production
4. Pricing at home and in export markets
5. Marketing and market determination
6. Transportation and related infrastructure
7. Refining and petrochemical industries
8. Utilisation of oil revenues: national, regional, and international development; regional and international flows
9. Integration of the oil sector with the national and regional economies
10. Arab oil and regional co-operation and complementarity
11. Arab oil and international relations: liberation; the New International Economic Order (NIEO)

The width and complexity of the policy areas listed, coupled with the desire to contain this book within rather narrow limits, make it mandatory that the treatment remain broad, away from technicality and detail, except where absolutely necessary.[4] Indeed, once it is recalled that policies are viewed here as broad guidelines for economic behaviour, the need to maintain consistently the discussion within general terms will become evident. It is also believed here that the adoption of this course will enhance a sharper focusing of the analysis, and a readier inference of its lessons and implications. With this in mind, we will now turn to a concise identification of the nature and content of the policy areas just listed.

*1. The Policy of Control.* In brief terms, what is meant here is the determination to take in hand the power of control over policy, or, stated differently, the adoption of the policy to formulate and implement policy. In most instances, this adoption was effected in the 1970s in defiance of the foreign concessionary oil companies. The first generation of agreements with these companies had been imposed by the erstwhile all-powerful, major Western concessionaires, often transnational conglomerates, and, by implication and projection, the governments behind them. The terms of these and subsequent agreements (or amendments and revisions) in effect continued to be enforced by outside will, down to the beginning of the 1970s, on the strength of the deterrent capability of the concessionaires. The question could be posed: How could the concessionaires impose their will on sovereign states? Does not the fact that government-company agreements were signed by both parties invalidate the claim that there had been an element of duress? This type of questioning is quite warranted, and it will occupy us later on in this study.

For the present, we will merely stress that the shift in the locus of policy — and decision-making — from foreign companies to national governments, is by far the most significant structural shift marking the 1970s, as was stated in the introductory paragraphs of this chapter. It is primarily a political/legal shift, one in power relations between the two parties. The significance of the shift in power relations does not lie merely in the dimensions of its political, legal, financial, and physical aspects, but also in its determining power with respect to the subsequent shifts occurring in all other policy areas which it made possible, and permeated thoroughly. The more detailed assessment of the import of the 'policy of control' deserves to be undertaken separately owing to its critical role. The next chapter will therefore be devoted to it.

*2. Exploration for Oil.* Looking for oil – prospecting, exploration, development with all the technical operations and skills involved – constitutes the first (pre-production) phase of upstream operations, assuming of course that legal and organisational prerequisite steps have been taken. These include the permission for exploration, the definition of the land area where exploration is to be undertaken, and the setting up of the institutional machinery that is to be involved in the operation.

The policy area under reference had witnessed a limited degree of involvement by national governments before the onset of the 1970s. This involvement had been mainly restricted to the legal aspects of exploration – the initial delimitation of concession areas and the subsequent narrowing of these areas or the cancellation of all or substantial parts of them – but acquired more substantial dimensions only in the 1970s. During this decade, exploration activities raised new issues for the national governments and became more closely associated with the question of reserves, depletion, and oil substitution, as we shall see in Chapter 3.

*3. Production and the Determination of the Volume of Production.* This policy area is very closely related to that of exploration; it is its offspring in fact. In those rare instances where a national government (via its ministry of oil, or national oil company) had taken over the activity of exploration, with all that that implied legally, technically, and financially, it naturally followed up with extraction, that is actual production. And, once production operations proceeded, it was up to the national authority to determine how much to produce. However, before the 1970s and under normal circumstances, there was little question of such determination of the volume of production.

This aspect of sovereignty was largely operative in one direction only: that of seeking expanded production to satisfy a thirsty foreign market, in a world still enjoying excessively cheap crude oil owing to the virtually complete design, control, and maintenance of the price structure by the major Western oil companies. The other, opposite direction, namely the restriction of production, whether for conservation, technical, economic or political purposes, was a feature of the 1970s essentially, and it will occupy our attention in subsequent discussion. Suffice it to say here that the conduct of production activities was the major policy area after that of pricing in which the principle of sovereignty over national resources came most frequently to clash with the principle of contractual prerogatives that the companies upheld, in the decades-old history of the oil industry before the 1970s.

However, a significant qualification ought to be introduced here. This is that the policies and measures of the 1970s relating to the disposal of accompanying gas, and to a certain extent of all types of gas, have not received the same intensity and effectiveness of attention of the competent national authorities, as oil has. But they certainly occupy a much more important place in national concern than under the concessionary regime. This is particularly true of the industrial uses of gas (as feedstock for the petrochemical industry), and in its liquefaction and export, in addition to its use as fuel for large-using activities like water desalination and electricity generation. This difference will receive evidence in the forthcoming examination of the production policies relating to oil and gas.

*4. Pricing at Home and in Export Markets.* The international pricing of oil is a most central issue for the oil-exporting countries. It is the determinant of the volume of revenue earned from exports, which has weighty implications for a wide range of issues, principally those of development and of conservation. On the other hand, the criticality of pricing for the importing countries is exceeded in significance only by that of the physical availability of oil. The determination of the declared (posted) price of a barrel of oil and the tax intake by the producing governments have been the issue over which these governments have had their most protracted, bitter, and inconclusive battles over the several decades of the history of the oil industry preceding the 1970s. Then, in one stroke, the exporting governments in October 1973 finally took in hand the decision to determine prices, and brought the seemingly interminable battle to an end.

The story of oil pricing has been vital to the exporting countries at least ever since in the early 1950s the basis of their share ceased to be a fixed sum per ton of oil exported and became a proportion of net profits. This is an oversimplification, since the governments' take consisted of a few components, though it is not intended to go into the detailed and long-drawn conflict over pricing and government take in this study, but only to discuss pricing policies as such and the foundations on which they were based. (Indeed, the two studies already referred to, by Ian Seymour and Fadhil Al-Chalabi, make it unnecessary for us to recapitulate the account of the conflict and the detailed points of disagreement. Though very recent, both can already be considered standard works on the subject, the first for its thorough and well-researched grasp of the details of pricing; the second for its broad and penetrating survey of the fundamental issues involved in pricing.)

Pricing for export markets is only one part of the story, though the major part. The other part is pricing for the domestic market. This is a policy area that has not received adequate attention so far, although its pertinence for supply conditions of hydrocarbons is becoming increasingly realised. The implications of present policies for the future in this regard will concern us in the forthcoming discussion.

*5. Marketing and Market Determination.* Since roughly 90 per cent of Arab oil is exported, the question of export and markets acquires special importance. Before the national governments took over the power of policy-formulation and decision-making, much of the argument between the proponents and the opponents of nationalisation, or of majority control by the governments, centered around the possibility (or the difficulty) of marketing for any authority but the major concessionary oil companies, with their network of marketing outlets and the absorptive capabilities of the network of their vertically and horizontally integrated operations. It was asserted by the opponents of national control that the governments of the producing countries might be able to extract the oil, with the installations already in place and in working condition, but that the marketing operation was quite another matter, well beyond the experience and capability of the governments. The successful handling of marketing by the latter is therefore a question of live interest, though admittedly a substantial part of the oil exported continues to be lifted by the former producing companies.

A part of the policy area of marketing is that of the allocation of specific volumes of crude to specific markets. In most instances, this involves no special problem, inasmuch as the process merely calls for pursuing the historic trend, with minor adjustments. But the question becomes much less easy to deal with if there is a substantial shortage in aggregate supply, in which case there will have to be a deliberate adjustment in the various shares of the customers. The question acquires more poignancy if a policy of deliberate discriminate rationing is to be adopted, involving some cuts in the sales to some habitual customers, or even an embargo altogether on certain sales. As a measure like this was taken by the Arab oil ministers in October 1973, it is worth examining within the context of marketing policy. (The measure was taken a day after the decision by the six Gulf members of OPEC, including Iran, to adjust oil prices by unilateral resolution rather than by negotiation.)

*6. Transportation and Related Infrastructure.* Implementation of the policy decision to take charge of production and marketing, logically involved acquiring and/or building the infrastructure for the gathering, storage, and transportation of the oil. A very large part of the infrastructural installations that were in existence by the end of the 1970s had already been built by the concessionaires by the beginning of the decade (pipe networks, gathering facilities, gas-separation facilities, storage tanks, loading terminals). But substantial additions were made by the governments during the decade, to account for the new productive wells drilled and developed, expanded production in need of loading or transport, new pipeline systems, and new terminals. More notably a great deal was done in connection with gas by way of liquefaction works and transportation and storage facilities. The substantial investment in tankers undertaken, both at the country level and at that of OAPEC, deserves recognition and examination because it touches on a much wider issue: that of the ownership of tanker fleets to transport the oil from the producers to the consumers.

*7. Refining and Petrochemical Industries.* Including these activities within the wide embrace of oil policy areas is essential since one of the major objectives of Arab oil producers is not to remain crude producers and exporters, but also to go into those activities that are advanced technically and rewarding financially. Of particular significance here is the marked contribution made by the development of refining and petrochemical industries to the cause of the more meaningful integration of the oil sector with the national economies (and the regional economy, for that matter).

It is necessary to indicate that refining and petrochemical industries are closely integrated activities, and that both are of great value to the process of development and to national security. However, the entry of the oil exporters into this dual-thrust activity is still hesitant and partial, compared both with their potential, and with the deep involvement of the international oil companies in refining and petrochemicals. This involvement was part of the fully-integrated system of operations of these companies, reaching from the oil well to the pump station handling gasoline distribution, on a worldwide scale. The Arab oil exporters have not gone into distribution outside their own countries, although most recently there have been some endeavours to enter that area of activity. A more forceful entry by the Arab oil countries into refining and petrochemicals, in their own territory and

*Identifying Major Policy Areas*

within the broader Arab region, but also in other regions of the world, raises policy issues which will receive consideration later.

*8. Utilisation of Oil Revenues.* The policy issues arising from and related to the uses of oil revenues are even more varied than these uses are. Broadly defined, the uses include public consumption (including welfare programmes and defence spending), development at home and abroad (the latter including aid for development within the Arab region and in other Third World regions), and regional security spending. They also include financial investments abroad as a residual item, in whatever form they are made. The policy issues relating to these uses are not only numerous, but also complicated.

The discussion of the policy area under consideration acquires particular significance owing to the pre-eminence of the policies comprised among the wide array of policy areas examined in the book. Indeed, the importance of the oil sector as a whole is fundamentally derived from the importance of the uses to which oil and gas in their various forms, and the revenues they bring about, can be put by and in the economies and societies of the exporting countries, the region, and the world at large. And, far from being merely economic in nature and scope, these uses relate to the performance, the well-being, and the future of the many millions of human beings they are supposed to serve. How, and how well this is achieved, will occupy the centre of the discussion of the policies to be examined in relation to the present section.

*9. Integration of the Oil Sector with the National and Regional Economies.* How closely is the oil sector truly integrated with the economies that form its habitat? This is a vital question that has to be answered, owing to the misconceptions attaching to the extent to which integration has been achieved. Integration will have to reach well beyond the mechanistic consideration of backward and forward linkages in the purely technical and obvious sense of the term, if the oil sector is to have a far-reaching impact. This impact it is entitled to have on the strength of the inherent physical importance of oil and gas (and their related processes and products), and the derivative importance of the hydrocarbons sector in terms of income generated, skills learned, social and economic organisation brought about, and international status acquired.

*10. Arab Oil and Regional Co-operation and Complementarity.* The vast development of the oil sector during the 1970s has several serious implications for Arab economic co-operation and for complementarity among the Arab economies. These implications are not one-directional, in the sense that they can move in a promotive or a prohibitive direction, depending on the orientations and policies of the main actors: the governments, the regional bodies, and the various business and professional communities. The positive value of complementarity makes it particularly essential to examine oil policies with respect to their contribution to co-operation and complementarity, and the extent and content of this contribution. This is all the more so owing to the significance of complementarity well beyond the boundaries of economics, particularly in the cultural, political, and security areas of the region's life and activity.

*11. Arab Oil and International Relations.* Three sectors within the area of international relations concern us in the present study: foreign aid made possible by the developments of the 1970s in the oil sector; promotion of the cause of the Palestinians for statehood in Palestine, and of the cause of liberation of Arab territories occupied by Israel; and the promotion of the New International Economic Order, NIEO, with its promise of certain central corrections which can redress the basic inefficiencies, injustices, and corrosive qualities of the present order. The first sector forms part of the concern of another policy area (number 8 above); the second will occupy us as we cannot ignore the place in Arab oil decisions of the cutbacks in production and exports to certain countries, and the imposition of an embargo on a few others, in October 1973. But the matter will occupy us only briefly, since it is contended here that the use of oil for purely political purposes is not a policy mainstream, but a decision that was (and could again be) only taken in extreme situations when the Arabs feel that some of their very central interests are seriously threatened. With the first and second sectors or sub-areas thus disposed of, the present, final policy area will focus mainly on the use of the instrumentality of oil in the promotion of the cause for a NIEO.

The eleven policy areas identified are not of the same significance to the oil industry or to its place nationally, regionally, or internationally. Nor has the struggle for the take-over of control over them been of equal priority and with equal determination. Indeed, a few policy areas were never contested by the oil companies, such as the pattern of oil revenue utilisation and the promotive role of oil in regional complementarity. For all these reasons, the policy areas will receive unequal attention and discussion in the present study, though all of them will be examined.

It ought to be pointed out that policy areas such as those introduced in this chapter are of significance for any kind of natural resource of important magnitude and role in the Third World, whose exploitation was initially undertaken by powerful foreign companies operating on an international scale. But they are of particular significance in the case of oil. Enough has been said about oil in the pertinent literature to make it unnecessary to dwell any longer here on its function and importance as a strategic commodity of primary and vital value in peace and war alike, for developed and developing countries, whether oil exporters or importers. In purely quantitative terms, oil is of paramount importance in the flow of international trade, ranking first in value of imports/ exports and representing about 15 and 18 per cent of total world trade for 1979 and 1980 respectively.[5]

But oil-exporting countries, while sharing with importing

countries experience of the physical uses and potential of oil, and of its centrality in their life both as a fuel and as a production input or industrial feedstock, stand almost alone in experiencing the direct impact and implications of oil exports with respect to a number of policy areas. These include the determination of the volume of production in relation to the reserves and the strength of the policy of resource conservation; and the receipt of sizable oil revenues and their utilisation for development, aid, defence, and economic and political leverage on the regional and international levels. It ought to be noted that the significance of oil for the Arab oil-exporting countries as a group is even greater, going beyond its place in the trade of these countries, where it accounts for 95 to 98 per cent of aggregate exports, to represent over half of Gross Domestic Product, GDP, on the average, a much larger proportion of foreign exchange earned, and a critical factor of immeasurable significance in the development process generally and in the industrialisation process specifically[6] — to say nothing of the special leverage it can provide in their pursuit of certain central objectives connected with the liberation of Palestine and the other Arab territories occupied by Israel.

It follows that a study of the power to formulate and implement policies relating to the control of a resource which is as important to the world at large as oil is, falls neither within the strict discipline and confines of technical economics alone, nor those of politics alone. It is, instead, a case *par excellence*, for treatment within the notional framework of political economy. Obvious as this statement may seem, the line of reasoning behind it has been largely refused and opposed for decades by the oil companies and a vocal sector in the developed countries behind them. The examination of the political-economy aspects of the control of oil resources will occupy us in the following chapter, where the shift in the locus of the power of policy-formulation, and the present process of policy-making, will be considered. This 'policy of control', as we have labelled it, is allowed a chapter by itself because of its determining nature and function in the other policy areas examined later in the book.

# 2

## THE POLICY OF CONTROL

### Backdrop to the 1970s

It would be quite wrong to say or imply that the Arab oil-exporting countries attempted to take over from the concessionary companies the power to formulate and implement oil policies only in the decade of the 1970s. There had been pressures and attempts before for the governments to obtain majority control of equity or to nationalise the concessionary companies' productive ventures, which is the strict sense in which the take-over of the policy of control is understood in this study.

Indeed, one concession agreement had stipulated equity participation as far back as 1925. This was the agreement with the Iraq Petroleum Company (then the Turkish Petroleum Company), which gave Iraq the option to buy up to 20 per cent of any issue of shares, 'Whenever an issue of shares is offered by the company to the general public'. But, as Seymour notes, this 'Machiavellian formulation' had a big catch in it, namely that the company 'was a closed private company whose shares were tightly held by a group of international majors ... and therefore did not, and indeed under its constitutional set-up could not, issue shares to the public'.[1] It was to be much later, in the decade 1948-57, that 20-25 per cent participation was to be offered to Saudi Arabia and Kuwait in the equity of companies seeking a concession to exploit the oilfields in the Neutral Zone shared equally by the two countries. (These were smaller American and Japanese 'independent' oil companies.) The last time participation was offered voluntarily, it was by a 'major'* company in 1961, when Shell accepted the principle of 20 per cent government shareholding for a concession offshore Kuwait.[2] (The Iraq Petroleum Company was in 1965 to agree to the formation of a joint company with the Iraq National Oil Company, INOC, in which INOC would have one-third interest, for the development of acreage outside the area of IPC operations. But the draft agreement was never implemented.) The appeals for outright, full nationalisation, were mostly made in the press and in successive Arab Petroleum Congresses, but did not result in a single instance of nationalisation before the 1970s.

Apart from the very few cases of minority participation effected, the control attempts by the Arab governments in the decades that constitute the backdrop to the 1970s were largely confined to price-related questions: the bases and rules that influenced the government's take per unit of crude exported, and determined the government's revenue from the export of crude, however calculated. The question of the volume of production was less frequently broached. On the other hand, the 1970s witnessed a vast widening of the policy areas which came to be controlled by the government, although, to be precise, price and volume determination remained the most important issues in their effect on all parties directly concerned. These parties, we should add here, were four: the exporting countries themselves, the developed, Western importing countries, the developing importing countries, and the oil companies — the erstwhile majors and the host of smaller ones, that is, the 'newcomers' or the 'independents'. (The Soviet Union, though a major producer and an exporter, was less directly affected than the other parties.)

*The 'majors' were initially seven companies (also called 'the Seven Sisters') — five US firms: Standard Oil of New Jersey (later Exxon), Texaco, Standard Oil of California (Socal), Mobil Oil, and Gulf Oil; the Royal Dutch/Shell Group (60 per cent Dutch and 40 per cent British); and British Petroleum (British). To these the Compagnie Française des Pétroles (French) came to be added. The smaller companies that entered the industry in the Middle East and North Africa beginning with the late 1940s and the 1950s were not associated with the 'majors'. Hence their designation as 'newcomers' or as 'independents'. Their number came to 15 or 16 by the 1970s.

Of relevance to this difference between the decades preceding the 1970s, and the 1970s, is the approach pursued or, one could say, the style adopted, by the producing governments seeking to take over control. In the early phase, the style was essentially one of reasoning or arguing with the companies to take greater account of the strong desire to have a larger say in the control and management of production and pricing, and of the pressing financial needs of the producing countries and their right, as real owners of the resource, to receive more revenue than they were in the practice of receiving. But there was no question that the power to determine the price of oil remained firmly in the hands of the companies, as well as the cost elements that together with the price, came on balance to determine the take of the producing government. The style of the governments, individually, and later collectively after the establishment of OPEC in September 1960, came to be somewhat firmer and more self-assured as Arab oil authorities acquired a somewhat stronger grasp of oil technology, economics, accounting, and marketing structures, and as political independence became more secure and inspired a stronger sense and tone of authority. These developments showed themselves increasingly as the 1960s progressed and finally faded into the 1970s.

Radical changes seen *ex post* rarely seem to occur all of a sudden. The movement towards the take-over of the power of policy-determination conforms to this generalisation, as the incursion into what the companies believed to be, and treated as their domain and area of exclusive competence, grew more determined. (It had been usual practice between the two World Wars for the concession agreements to grant the oil companies 'the exclusive right to explore, prospect, drill for, extract, treat, manufacture, transport, deal with, carry away and export' oil and other hydrocarbons within the area of the concession.[3]) Entrenched in their financial, technological, logistical, and institutional strength, and enjoying their world-wide networks of horizontal and vertical integration, the majors made full use of the exclusiveness of their rights and their multi-sided capabilities. It is no wonder, therefore, that the 'incursions' made into the companies' area of authority were gradual and small-scale, even late into the 1960s. Furthermore, as subsequent discussion will show, in the vast majority of cases, tangible adjustments or revisions in agreements only occured when the companies realised that the governments, in exasperation, were on the point of introducing unilateral legislation to force the companies to cede what had been in fruitless debate and negotiation for long periods of time.[4]

It was the 1970s that were to witness landmarks where major and radical shifts in the power relationship took place. These landmarks we will identify later. For the moment, suffice it to say that the shift in the locus of power characterising the 1970s was direct, identifiable, and sudden, associated as it was with certain direct, identifiable, and sudden events, decisions, and dates. In turn, the suddenness and huge scale of the shift came thanks to a drastic change in approach and style by the governments. This was based no doubt on the change in the power structure, owing to juridical measures taken relating to majority government participation in company equity, or to outright nationalisation. One can sum up the change in the approach and style in that the approach had earlier been shy, irresolute, and essentially dependent on the claim of the justice and logic of the governments' position, but that it had changed into one which became firm and self-assured enough to enable the governments to transform the conceived justice and legality of their position into policy decisions: that is, essentially into political decisions.

This shift in approach, and the utilisation of the instruments of sovereignty after the course of persuasiveness had failed, became possible thanks to three other differences between the phases of the pre-1970s and the 1970s. The first was the growing political power of the oil governments in an international climate which acknowledged and promoted the power of Third World countries. In turn, this was bolstered by three developments: greater balance between the two superpowers, the USA and the USSR; the receding power of the European colonial states; and the growing weight of the United Nations forum and its moral support, even if it remained of limited effect. At the same time, the second, economic difference between the 1970s and the decades forming its watershed made its full effect felt by the end of the 1960s.

One component of this factor was the gradual erosion of the exclusive control by the majors of the oil industry at the international level, owing to the increasing entry into the industry of smaller but none the less cumulatively influential oil companies, the 'newcomers' or 'independents'.* These newcomers were willing to offer the governments better terms than the majors (including equity participation in some cases, and more liberal profit-sharing arrangements in some other cases), in order to acquire a footing in the international oil industry. The second component to consider was the tightening of the oil market with the turn of the 1970s, after the soft 1960s which actually witnessed a drop in the absolute level of current prices. The two components together gave the oil countries considerable bargaining leverage with the majors and, by extension, with the independents as well.

But the third development was probably the most significant and powerful in the process of change in power relations. This was the emergence of OPEC, and the desire and growing ability of its members to act in concert (at least, initially, with respect to oil pricing, costing, and accounting). The appearance of OPEC came at an appropriate conjuncture when the two developments just referred to began to interact in a manner that urgently called for, and justified, corrective action. By the time the experience of OPEC had matured sufficiently to enable it to take firmer and more far-reaching action, the decade of the 1960s was fading into the 1970s, and the stage was set for the major act of take-over of control. The cumulative effect of the various factors indicated led to the qualitative difference between the long era preceding the

*Most of the 15-odd newcomers were American, with a few West Europeans, one Brazilian, and one Japanese company.

1970s, and that of the 1970s. This was the much greater effectiveness of the action of the 1970s culminating in the take-over of the power of policy-formulation, compared with that of pre-1970 attempts, which had been marginal in scale and very limited in scope — in one word, undistinguished.

## Why Policy Control?

At this point, an uninvolved observer might ask: Why is the 'policy of control', or the take-over by the governments of the oil-producing countries of the power of policy-formulation and implementation so important for them? Why not leave the whole oil industry in the hands of the international oil companies to handle, given their expertise, their long experience, the substantial financial resources at their disposal, and their vast integrated network covering all oil operations, beginning with the one furthest back upstream, and ending with the one most distant downstream? What have the oil countries got to complain about strongly as to want to run the oil industry themselves? This type of question is not rhetorical and hypothetical; nor is it posed only by the uninformed. For instance, it lies behind statements like the one quoted below which appears in the Preface of a well-known book by Longrigg, a respected historian of the oil industry in the Middle East:

> Of the third element, the foreign concessionaire companies to whom such rights have been granted by the governments concerned, some readers who are convinced in advance of the greed, the intrigues, and the oppressions of Big Business may feel that too favourable a view is taken in these pages. But a favourable view seems to the writer to be imposed by the patent facts of the case: whatever evils of callous exploitation may have existed in other regions of the world or in other periods of time, the great oil companies operating in the Middle East in this century do not seem to have exemplified them. Impartial scrutiny

seems to show, on the contrary, that with whatever motives of self-interest or of genuine benevolence, and with every advantage of great commercial prosperity and abundant resources, the British and American companies have been directed and served in their Middle Eastern dealings by enlightened men of goodwill: that their record while falling short of perfection, will bear critical inspection: that they have performed great services and few disservices to the countries where they have worked.[5]

It would indeed be difficult to reconcile this idealised portrait of the oil companies and their leaders with the natural tendency of company leaders to make full use of the 'exclusive rights' to which their agreements with the producing countries entitled them, a sample of which we quoted a few paragraphs above. Given, in addition, the mastery by the companies of the world market for oil at all its stages and in all its forms; the weakness of the oil countries and their non-readiness to undertake oil-related operations and activities (whether in terms of organisation, networks, technology, or finance); and the decisive support which the Western governments gave to the oil companies — given these facts, it would seem to us that Seymour's assessment is much more apposite than Longrigg's:[6]

Particularly in the quarter of a century, which followed the Second World War, this freedom to decide upon output levels in the various producing countries was a source of enormous strength for the major multinational oil companies with their high degree of horizontal integration — which means to say that, severally and collectively, they owned and controlled a wide variety of crude oil sources across the globe. This, in conjunction with the interlocking pattern of ownership by the majors of the various operational consortia in the producing areas, conferred on the companies the means to co-ordinate and plan the bulk of non-communist world supply entirely in relation to the market requirements of the industrialized

West and the convenience (technical, financial and logistic) of their own integrated systems, without necessarily showing much regard for the particular needs of the countries where the oil was produced.

For the producer governments, on the other hand, the privation of the power of decision-making on oil production levels, as well as on pricing and management of operations, was tantamount to a loss of sovereignty in an extremely vital area of the life of their countries. It was as if an essential steering mechanism was missing from the ship of state. It was difficult, if not well-nigh impossible, to plan for the overall development of a country's economy so long as the massively dominant oil sector remained a self-contained enclave, a state within a state, subservient to the dictates of international rather than national exigencies.

The fact that the last statement was written in the context of a discussion of the control of production, and not control of operations in general, adds to its critical impact rather than reduces it. Furthermore, though the quotation comes from a very recent book (dated 1980), this does not mean that similar feelings did not exist in strength much earlier. An article published as far back as January 1957 takes issue with the companies for more or less the same reasons, in spite of the distance of about a quarter of a century between the two quotations. The older article said, among other things:

> The place of the oil industry in the social and economic organization of the oil countries is another important, and often overlooked, aspect of Arab oil. The increase in revenue — both total and per ton — has lessened but has not removed the 'alien-ness' of the oil industry. Even the utilization of oil revenue for development purposes, as in Iraq and to some extent elsewhere, has failed to remove completely the feeling that the oil companies are 'outsiders.' Here lies a very sore spot in the relations of the companies with the oil countries. So long as these countries feel that

the oil industry involves operations beyond their skills and control, and so long as they feel that they are no more than the titular owners of oil or its transit depot, they will not be able to feel that the oil industry is an integral part of their society and their economy. It is beside the point to claim that since these countries accept the payment offered for their oil they ought to stop complaining. For the issue is not a legal one, or even an economic one. The oil industry will have to become part of the performance of the economy and to merge as far as possible with the country's social and economic organization if it is to be freed from the onus of 'outsideness.'[7]

Nor was the companies' attitude of hostility and refusal towards any desire for participation in decision-making by the governments restricted to their dealings with 'hawkish' governments (or radical governments, in the current parlance) such as Libya, Algeria and Iraq. We encounter the same attitude in reaction even to Saudi Arabia, where the climate of government-company relations has for many long years been the most friendly and characterised by 'mutual understanding'. Thus, when Saudi Arabia indicated its intention in 1965 to join an effort by OPEC to undertake production programming involving the setting of production shares for OPEC members out of the total production volume considered appropriate for the whole organisation, Aramco, the producing company, rose in protest and threatened to take the government to international arbitration. It is instructive to add here, that the programming envisaged was still a tentative, shyly approached subject.

The case for the take-over of decision-making, from the Arab countries' point of view, can be stated both negatively and positively. In its first form, it proceeds to stress the unfavourable terms which the various oil countries had to accept when the concessions were first granted to the major oil companies.[8] In most instances, this took place when the countries involved were still under foreign domination, whether overt and formal, or de facto but disguised behind formal independence. The vast discrepancy in power relations between the European powers and the United States to which the majors belonged, and the oil-producing countries, made the granting of the concessions on the terms then prevailing, virtually 'contracting under duress'.

The claim that the terms were not particularly generous because oil prices themselves were low, is really circular reasoning, inasmuch as the majors themselves set the level of prices to suit their integrated operations, since they were in fact buying oil from themselves, given the 'kinship relationships' between the producing companies and their affiliates in the subsequent phases of refining and distribution. Further, to the political/military weakness of the owners of oil must be added their inability, technologically, financially, and managerially, to develop their oil, extract it, and market it — let alone to refine, transport and distribute it. Agreements subsequent to the first generation (whether altogether new agreements, or amendments and revisions) were handicapped by the low level of the base on which they were built, and could not, therefore, contain anything more than very marginal improvements — ones which were usually so small as not to constitute noticeable cosmetic embellishments. The sceptic has only to read the history of the long and arduous process of negotiation to bring about these improvements, down to the end of the 1960s, to be convinced of the point being made.

It is useful in this context to remember that the producer countries in the early phase of the concession system, namely from the mid-1920s to the end of the 1940s, by and large obtained a mere 4 shillings (gold) per ton produced. This system was replaced at the turn of the decade of the 1950s by the '50-50 profit-sharing formula'. Government's take per unit of export improved subsequently, but it none the less remained small for three reasons. These were the low price at which oil was sold or 'posted', when it hovered around $2 per barrel; the deductions made from gross sales (selling discounts, etc.); and the system of accounting adopted in the calculation of net profit, involving the deduction from net profit accruing to government of the royalty to be paid (in lieu of 'economic rent') instead of 'expensing' it, that is, including it among the costs of operation before arriving at the figure of net profit.

Even after protracted debate and negotiation finally provided some satisfaction with respect to the costing of oil and the expensing of royalty,[9] the low level of the posted prices (which in the 1960s were even higher than the market prices in effect obtained) left the producer countries with meagre revenues to show against the sale of considerable quantities of oil and the depletion of a most precious, non-renewable resource. This eating up of national capital was a most serious matter. For, as can be readily realised, the governments were in effect giving up a valuable asset which is stored in their sub-soil at no expense, for financial assets which are subject to value depreciation, official devaluation, and price inflation. (The danger of the freezing of any surpluses belonging to the oil producers and placed in Western money markets, like very steep inflation, was to come much later — that is, beginning with the early 1970s, when there was enough revenue to lead to the accumulation of surpluses.) In rigorous economic terms, the oil revenues should not be considered a component of current national product, since they are not a flow deriving from an activity based on the production and sale of a renewable product, like timber, wheat, washing machines, or electrical generators.

The case for the take-over is even more compelling in its positive form. It starts with the need to extend the principle of sovereignty to cover natural resources. Admittedly, the desire for extension, in part, has political and socio-psychological justifications. But this is not all. In larger part it has economic justifications, considering the importance of oil resources and the revenues they can generate for economic and social development, (including the acquisition of new skills and the introduction of new oil-related activities and industries), for public services, and for national defence (the last objective to be understood as having non-political and non-military aspects and implications as well).

Another compelling reason why the oil-producing governments finally decided on the policy of control, was because only after an effective take-over could they design all the component policies and activities that fall within the framework of the oil industry broadly defined, with internal consistency among the components assured, in harmony with their national, regional, and international policies – economic, political, and strategic. Thus, the adoption of the principle and objective of oil conservation could only be pursued through the formulation of an appropriate mix of policies relating to pricing, volume determination, and revenue utilisation (the last including development and foreign aid). Likewise, the determination of the volume produced and marketed, and the price charged, could only be effected after due identification and careful balancing of national interests and regional and international responsibility. Again, volume-determination has to be further harmonised with the downstream needs of the oil countries, and these are a function of their refining and industrialisation policies. Finally, the volume-cum-pricing policies are also a function of the producers' estimation of their overall financial requirements and of their attitude towards the building, or alternatively the avoidance, of surpluses abroad.

Although there is no need to present other illustrations in support of the necessity to integrate different policy areas and to harmonize the policies, strategies, programmes, and measures within them, it would be instructive to add just one more argument in favour of the policy of control. This is the argument of the *ex post* lessons of experience: the difference that the take-over has meant for the producing countries, in terms of revenues, economic power, political leverage, and the satisfaction of the desire to extend national sovereignty to the realm of natural resources.[10] In other words, this is the argument that proceeds by asking: What would have happened had the take-over not occurred and, instead, had there been mere improvements in favour of the national governments in the terms under which the foreign oil companies operated? To answer this kind of question, one has to survey the situation at the end of the 1960s, before the critical decade of the 1970s had arrived, and to assume an acceleration of the accrual of benefits to the governments, in order not to prejudice the answer unfavourably.

No radical transformation can legitimately be assumed, within the structure of the contractual framework then in existence, but only modest improvements in the terms, even if we allow for a speeding up of the rate of improvement. Judging by the experience of the 1950s and the 1960s, the current price had hardly changed in 20 years; it had even dropped slightly in the 1960s. If we were to assume a reversal of this trend owing to the tightening of the market, but continued determination of the price by the companies with pressures by the governments for some increases, most probably it could not be conceived that the price would have exceeded, say, $4 per barrel, by the end of 1979. This would have meant just over a doubling of the price prevailing at the beginning of 1970 — a vast improvement over the record of the 1950s and the 1960s. Against this scenario, we find that, in fact, the marker crude Saudi Arabian Light 34° was raised to $5.119/b in the unilateral adjustment of prices in October 1973, moved to $11.651/b on the first day of January 1974, and ended by reaching $26/b on the first day of January 1980. The comparable crudes of other Gulf producers were higher; those of the Mediterranean producers, Libya and Algeria, were over $8/b higher by the beginning of 1980.*

The vast improvement in government take, important as it is, does not tell the whole story, and it is necessary here to refer also to the impact of price adjustments since October 1973 on the search for other sources of hydrocarbons than those known or developed at the time, and for other sources of energy than hydrocarbons — in addition to the new opportunity which the upward adjustments provided for the exploitation of formerly sub-marginal oil reservoirs. It is further necessary to note here that the producing countries, without the take-over, would have foregone — in large part or wholly — the other benefits that came to be theirs in the

*All quotations above are at current prices (not adjusted for inflation).

various policy areas: the opportunity to undertake exploration for oil themselves and enjoy the experience that can be gained thereby, the promotion of the building of upstream infrastructures, the establishment of refining and petrochemical industries, the building of fleets of tankers, the training in skills and institutional set-ups not available before, but above all the launching of various important developmental programmes which became possible only with the inflow of substantial oil revenues that could not have accrued under the old system of control, and of the important economic and political leverage in the international field which came to be within the governments' grasp as a by-product of the take-over.

In brief, opportunities foregone without the take-over are no less important than advantages reaped with it: both have to be included in the profit-and-loss account. Even without 'over-proving' the substantial net advantages of the policy of control by further pointing to the possibility of finding fault with the policies and behaviour of the oil companies during their era of control on economic, political, and even moral grounds, the case for the policy of national control is — we contend — firmly made and needs no additional evidence. We also believe that the case is equally made that the companies did not bestow the kind of blessings on the producing countries that is implied in the quotation from Longrigg which we reproduced earlier.

### Why Not Earlier Policy Control?

If the take-over of control was as necessary and useful to the oil governments as we set out to show in the last section, why had it not been undertaken before the 1970s? The question acquires greater relevance if it is realised that the benefits were not restricted to the oil-exporting countries, but were world-wide in scope, including in their reach the fields of energy as a whole, the industrial consuming countries, and the Third World. (The reader is requested just to tolerate this assertion for the time being; the rest of the book will attempt to prove it.)

A strong current can be detected all through the discussion in reply to the question with which we have just started, but can be felt more strongly in the unfolding of the history of the oil industry. This is that the degree and quality of authority and control which the Arab oil governments actually achieved and exercised at any one moment was the resultant of the interplay of promotive pressures acting on their side (that is, in favour of take-over), and deterring constraints acting against them (that is, in favour of the continued control by the foreign oil companies). The pressures were generated by an aggregation of factors: the political desire for control; the economic need for more oil revenues from a unit of oil extracted, in compensation for the depletion of a precious non-renewable resource; the urge to acquire the technology involved and to train nationals and equip institutions in the various skills called for in the many aspects and phases of the oil industry; the need to upgrade the organisational structure of government and the economy in order to make it capable of gradually running the industry; and, as an overall desideratum, the urge to bring about comprehensive, self-sustaining social and economic development, and an acceptable degree of national security to protect the nation's interests and well-being.

On the other hand, the constraints consisted of the power of the companies to resist the expression and the satisfaction of the pressures, with all the forces that this power represented and generated in the economic/financial, technological, institutional, legal, and political fields, and the public relations and information output it could use in the service of the image, policies, and operations of the companies, particularly in the Western industrial countries. It need hardly be added that the companies started with an advantage in these countries, as they enjoyed overwhelming sympathy there which the oil governments did not.

It would have been practically impossible for the exporting countries to segregate and weaken the bases and aspects of the economic, technological, institutional, legal, logistical, and political power of the majors, inasmuch as these aspects were inter-related and supported each other in both directions. This is to say that the companies in the first place acquired their power and later solidified it because of their economic, technological, and other advantages; and that their power, once acquired, permitted more or continued economic, technological, and other advantages. The dimensions of their world-wide networks (the horizontal aspects of integration), and the tightness of the chain of successive operations which they owned and controlled, upstream and downstream (the vertical aspects of integration), plus the fact that they co-ordinated their operations particularly with respect to exploration, production, pricing, and marketing, gave the 'Seven Sisters' vast and nearly impregnable power *vis-à-vis* the producing countries. They owned and marketed all the crude available outside the centrally-planned countries, and owned the bulk of the refining, tankerage, and distribution facilities needed for the crude in hand. This made the majors a tight oligopoly which was different from the standard model of oligopolies, in that the associates had a binding working arrangement. This may not have always been explicit or discernible, owing to Anti-Trust regulations in the United States, but there was certainly a *de facto* collective understanding to govern their production, pricing, and marketing policies.[11]

For over three decades, that is, till the late 1960s, the majors were able to have and enjoy the sources of power to which reference has been made thanks essentially to the legal factor, that is, to the concessions and other contractual arrangements in their hands, and the security they felt as a result. But, as indicated above, they also had all the other essentials of power – finance, technological supremacy, organisation, appropriate facilities, in addition to political support by their governments. The combined weight of all these components of power gave the companies the added power to intimidate and later effectively to influence the governments of the oil-producing countries. This situation remained largely unchallenged till the mid-1950s, or over 30 years after the first agreement had been signed with an Arab country, that is with Iraq in 1925.

On the other side of the power balance (or, more aptly, the power imbalance) lay the producing governments. These were weaker than the company group on every single score. Their signature and seal had been affixed to concessions and other agreements with the oil companies, though in fact this was done in the first place under conditions of political subordination to the Western governments under whose patronage and protection the companies operated. And the penalty of withdrawing their signature through legal and political unilateral action would have been overwhelming, as Iran's Mossadegh had found to his own and his country's chagrin.[12] The economic and financial penalty would have been no less daunting, owing to the international reach of the grip of the majors on sources and markets of crude, and the tight oligopolistic network which held the companies together and deterred any breakaway movement. (They had every economic reason to stay together, anyway.)

The economic and financial resources of the producing countries were on the whole so meagre that the oil revenues, though awfully modest, were a welcome addition. But, in any case, their technological capabilities and practical experience were too slim to enable them to take over the production, transportation, and marketing of the oil – let alone the production of refined products, as the refineries were company-owned, within the producing as well as the consuming countries. And, above all, the countries were not joined by any *de jure* or even *de facto* institutional arrangement and could only act individually, if at all.

This was indeed a vicious circle. The many-sided power of the concessionary companies gave them the exclusive opportunity to produce, price, carry, refine, and market oil. This power and the foundations on which it rested made it virtually impossible for the governments of the oil-producing countries to make incursions into the three most strategic sectors of the companies' power: production, pricing, and marketing. This in turn strengthened the position of the companies and permitted them to remain essentially unchallenged before the 1960s, and thereafter in the 1960s only with partial effectiveness — that is, until the fragments of a cumulative process of countervailing power build-up permitted a breakthrough in the iron circle.

The breakthrough did not come in any one stroke or dramatic development, but, as stated earlier, in a series of discrete steps and processes. Four of these stand out and should be stressed, in the context of the enquiry as to why the take-over of control did not and indeed could not be achieved much before the 1970s. The *first* was unrelated organically to the remaining three, although it made the unfolding of these easier. It was the emergence of independent oil companies, the newcomers, which were not members of the giants' grouping, the seven (or eight) major concessionary companies. Ironically, it was the United States government which encouraged these independents to go into oil exploration and production overseas. It would be interesting to speculate, but one would probably never know for certain, whether this government would have pushed its policy of encouragement of incursion into the field of the majors, had it then realised that this would ultimately help undermine the power base of all oil companies, majors and smaller independents alike.

The incursion was performed effectively through a combination of three factors: the availability of acreage, or plots of land in the Arab producing countries, with promise of oil finds, not assigned by concession, or originally assigned but subsequently reclaimed or relinquished owing to the failure to explore there; the offer of better terms to the oil governments in order to gain the coveted foothold; and the political support by the United States in their eagerness not to leave much of the oil industry in the Arab countries (in the Mashreq as well as the Maghreb\*) in the hands of West

---

\*The 'Mashreq' region includes Egypt as well as the Arab countries in Asia, while the 'Maghreb' includes Morocco, Algeria, Tunisia, and Libya. Mauritania, Somalia, and Djibouti seem to be left out from the second category.

European countries and companies. (The exceptions to this last generalisation were in Saudi Arabia, where the concessionary companies were all American, and Kuwait, where American companies were half-owners.[13])

The *second* factor was the build-up of strong resentment against the oil companies in the producing countries, owing to the combination of certain elements. These included: the central importance of oil already perceived; the complete control of the oil sector by the concessionaires in a way that made it a truly 'enclave sector'; the alienness of these concessionaires as perceived by the countries;[14] the interference in internal affairs by the concessionaires as claimed by articulate groups of intellectuals, journalists, and politicians; the unsatisfactory exploration and/or production policies and rates pursued by the companies; and dissatisfaction with prices charged and revenues earned. Resentment against the West in general, including the United States, was motivated further by a special factor, beginning with the 1940s: Western support of Zionism, and later Israel when the Zionists succeeded in setting up a state in Palestine in May 1948 and in uprooting more than half its Arab population and forcing them into refuge in the neighbouring Arab countries. This support was reflected in anger, dismay, and frustration which, together, found their objects in the West, mainly Great Britain and the United States, but also France, West Germany, Holland, Canada, and Denmark.

Pressure at the levels of government and the public did not turn into explosive gas, and did not lead to major outbursts and confrontations between the oil-producing countries and the oil companies. However, it is instructive to note here that on three occasions when the outbursts resulted in a disruption of oil supplies, this occurred as a form of confrontation with the West over the issue of Palestine, or Palestine-related issues. These occasions were the Suez war of October 1956 launched by Britain and France, with Israel in collusion,[15] against Egypt, when the flow of Iraqi oil to its Mediterranean outlets was stopped in Syria; the war of June 1967 when, in addition to disruption of flow, there was a partial embargo on exports to the United States, Britain, and (by some exporters) West Germany; and the October 1973 war when there was a longer and much more elaborate combination of general, graduated production cut-backs with a selective embargo (the latter involving the United States, Holland, and Portugal; South Africa is continuously embargoed owing to its apartheid policies). Apart from these occasions, the pressure was economic in motivation, and self-controlled in expression (some would say excessively so). But it slowly produced the effect of identifying and crystallising the grievances of the producing countries, and pointing to the outlets of protest. These began by being very narrow and limited, as we indicated earlier, but widened as the four developments we are in the process of discussing unfolded.

One of the constraints that reined in the Arab expression of discontent and protest was the feeling that the producers were not yet prepared in the various skills called for by the production and marketing of oil, as well as by the harmonisation between supply quantities and sources with demand volumes and outlets, on a global basis. On the other hand, there was ambivalent feeling with respect to the facilities needed in oil operations. For, while the advocates of the nationalisation of oil saw no problem inasmuch as the facilities were already there, and what was needed was just a political/legal step to take the concessionary companies over and thus come to own the facilities, marketing posed a much tougher problem which the opponents of nationalisation naturally emphasised (or even exaggerated). Even the advocates of nationalisation did not underestimate the problems associated with marketing, though they did not consider them insurmountable. Thinking about the issues of control of production and marketing pointed to the two courses capable of leading to solution, namely equity participation and nationalisation, though this did not then provide the actual acceptance on any scale of nationalisation, except in theory.

The shortage of skills and experience was to be met by technical, managerial, and financial/accounting training, as well as by participation — even if on a small scale and gradually — in equity and higher-level management. Consequently, the first long-term process of redress was to be the training of nationals in the various skills needed for the operation of the industry, but, to be precise, largely in production and the building and maintenance of the infrastructure for upstream operations, down to the export terminal. Iraq was the country that took the pioneering, most energetic, and most effective steps in this direction, followed by Algeria. Its head start was achieved both by design and because oil operations on a large scale for export had been started there before those in any other Arab country.

The problem of marketing was essentially one of market structure and politics. Any one producing country attempting then to place its oil on the market in defiance of the group of oil companies would have failed. This was partly because there was then no one country which stood way ahead of all the other producers with respect to the volume of production, and which therefore could pioneer in independent marketing without fear of blockage by the international company network. (Saudi Arabia's pre-eminence as a producer was to come much later, in the 1970s.) The failure would also have been caused by the fact that the market was run like a 'closed shop': all entrants had to be members of the union of the Big, so to speak. And the union was the Seven (or eight) Sisters. In brief, no one producing country *by itself* could stand up to the oil companies and defy them successfully, whether on the issue of marketing, pricing, refining, or other basic oil-related policies and activities.

The cure which suggested itself for this source of serious weakness was the grouping of the oil countries, as a countervailing power to that of the companies in close association. Hence the creation of the Organisation of Petroleum Exporting Countries, OPEC, in September 1960, in a meeting in Baghdad of the five founding members (three Arab: Iraq, Kuwait, and Saudi Arabia; two non-Arab: Iran and Venezuela). This was the *third* form of redress in the group of four forms which we set out earlier to identify and examine. However, as the discussion has suggested, OPEC was not designed to be, and did not turn itself into, a mirror image of the oil group, or cartel as many prefer to call the Seven Sisters. Its objectives, and subsequent achievements even to this day, fall far short of those undertaken by the oil companies when they formed a tight association of decision-makers which designed and implemented production, pricing, and marketing policies in harmony and solidarity. OPEC has never gone beyond the much more limited objective of co-ordinating its position and approach with regard to the questions of pricing, costing, tax, and later participation. Nevertheless, in spite of this serious limitation, OPEC's emergence has been an historical event and a prominent landmark in the experience of the oil exporters — Arab and non-Arab alike. (In fact, it occupies a position of pre-eminence, almost uniqueness, among all Third World associations dealing with the production and sale of raw materials and natural resources.)

The *fourth* and last factor to examine among the factors which came together to bring about the breakthrough in the vicious circle to which we referred earlier, was the daring and defiance of a small number of oil exporters, in their confrontation with the oil companies. It was stated earlier that Iran's sad and costly experience of abortive nationalisation over the years 1951-3 constituted a very strong deterrent to those countries wishing to challenge and loosen the grip of the oil companies on all major decisions of the oil industry. However, three Arab countries subsequently distinguished themselves in the decisions they took involving the recapture of a degree of control in certain areas of the industry. (In fairness, it must be said that Venezuela played a very prominent and decisive role of self-assertiveness and defiance before and during the formation of OPEC, but its role is left out of the present account which is restricted to the Arab oil exporters.)

The three were Iraq, Algeria, and Libya. In fact, the policies of these countries still retain a distinctive flavour of independence and defiance today. However, as their efforts and those of their fellow-members of OPEC to improve the situation of the industry during the 1960s, which constituted a period of soft markets and low prices, only crystallised and bore fruit in the 1970s, they will occupy us in the next chapter which deals with policies relating to upstream and downstream operations. For the moment, it is sufficient to point to some of the measures taken by these three countries before the 1970s, which were to influence greatly the course of events in subsequent years, within the confines of 'the policy of control'.

By the turn of the decade of the 1960s, Iraq's grievances against the oil companies had grown into a long list.

> The issues involved included: equity participation, profit-sharing arrangements, relinquishment of unexploited acreage, natural gas utilisation, cost components deductible from gross sales, posted price determination, marketing allowances, appointment of an Iraqi managing director, Iraqisation of staff, use of Iraqi tankers... But essentially the controversy was the result of Iraqi unease over the fact that the oil was being exploited by a foreign concern, not the country itself. Thus the issues have always been at the base political-ideological, relating to the desire to have complete mastery of one's resources and to put these in the service of national ends.[16]

The government's reaction to the unsolved issues was unilateral action, in the form of legislation. Decree 80 of December 1961 reclaimed more than 99.5 per cent of the concession area of the Iraq Petroleum Company, and left to it only the acreage actually under exploitation.

This was the first major official step in a long course of confrontation, which ended in June 1972 with the nationalisation of the IPC (except for the French share; however, this share and the Basrah Petroleum Company and the Mosul Petroleum Company were nationalised subsequently). The confrontation led to a considerable slowdown in exploration and production by the oil companies in the 1960s. But it stiffened the determination of government, backed by public opinion, to be patient with the oil revenues that were growing less fast than elsewhere in the Arab oil countries, in spite of the urgent need for these revenues for the ambitious and expanding development plans formulated, particularly from the second half of the 1960s onwards.[17]

The attitude predominating in the 1960s led, a decade later, to the take-over of IPC by nationalisation and the assumption of the various aspects of sovereignty over the oil resource. Furthermore, the slowdown in production, which meant at the time the foregoing of a certain amount of revenue (estimated considerably differently by different writers, in a range stretching from $221 million to $1,265 million)[18] must now look like a blessing with the oil price up from, say, under $2 per barrel on the average, to $34 per barrel by early 1982, with the added difference that Iraq received less than $1 per barrel (half the net selling price), whereas it currently receives the full selling price minus the tiny cost of production and any allowances made to the companies lifting the oil under long-term contracts.

Algeria's struggle for control also started early. Before independence in July 1962, the French had granted concessions for exploration and production to French companies. In the Evian Agreement of March 1962, which governed the final political settlement, the French had insisted on the retention of the southern desert under their control, mainly to exploit the rich mineral resources there, including hydrocarbons. But subsequently, after costly struggle, the Algerians succeeded in defeating partition and having their sovereignty undivided over the whole country.

Three years after independence, the Algerian government had formulated a framework to govern French-Algerian oil relations (codified in an agreement on 29 July 1965). This involved, among other things, the abolition of the concession system and its replacement by one of co-operation within an 'Association coopérative', under which plots were allotted to prospectors for exploration and development; the take-over of all operations involving the treatment and domestic distribution of gas produced jointly with oil, and of all gas destined for export except that sent to the French market (left under a joint Algerian-French company); and the amendment of the profit-sharing formula, from 50-50 to 55-45 for the government and the companies respectively, with the stipulation of a base price for profit calculation. Algeria took further steps which enabled it to obtain effective control over its oil and gas in the span of about six years, between the agreement of 29 July 1965 and the nationalisation decrees it enacted in February 1971. These were followed by the issuance of a new oil code in April 1971 which served as a framework for the development of the industry in its various phases. However, the legal measures taken in the 1960s stopped short of controlling production, while placing the other aspects of the oil industry under Algerian control and operation. The degree of control of production was 17.75 per cent by the end of 1969, but it rose to 35, 56, and 77 per cent for 1970, 1971, and 1977 respectively.[19]

Libya, the third country considered in the present context, was a late-comer to the industry. But,

> as has been rightly observed by a leading Libyan economist, Libya was fortunate that oil was discovered in the late fifties, and not a decade or more earlier. Had it been discovered before independence [which took place in December 1951], the European countries then controlling the destiny of Libya would have manoeuvred to block independence and would have clung to the resources by all kinds of delaying tactics. Furthermore, the lesson of Iran's nationalization of the Anglo-Iranian Oil Company in 1951, and how the company succeeded — with the political support of Britain — in bringing the oil industry to a standstill, and thus penalizing Iran while the controversy lasted, was of great value to the Libyan authorities. This made them decide on a policy under which no one company would be able to 'blackmail' the country or immobilize its oil industry.[20]

Not only was this policy valuable for Libya and for the purpose suggested, but it was also valuable for the whole community of oil exporters, inside OPEC and outside it, in that the large number of independent oil companies, the newcomers, who were allowed to explore for oil in Libya more than anywhere else, made a split in the tight network that the majors had created and sheltered over the decades, permitted more competitiveness among oil companies, and in the long-run gave the oil countries greater bargaining leverage. This is true although the activities of the newcomers led to the lowering of oil prices and thus created a strain on the producing countries.

The main landmarks in the drive for control of the oil industry stand in the 1970s, and will be examined in the following section of this chapter. But it is fair to give credit to the new Libyan government which took over in September 1969, for deciding on a 'policy of control' soon after the assumption of power. The new government's resoluteness in this respect was to be effectively translated into a 'sustained policy of growing control'[21] first in the area of price determination, and subsequently in other policy areas as well.

The singling out of the three countries, Iraq, Algeria and Libya, does not mean that the other Arab members of OPEC (or, for that matter, the non-Arab members, particularly Venezuela) had not been active in improving and solidifying their position *vis-à-vis* the oil companies in a desire to obtain control. It is instead meant to suggest that the three pioneers were not only convinced that nationalisation, whether full or partial (under some formula of equity participation), was the only practical way to get equitable terms for the oil extracted, but went ahead preparing themselves for nationalisation. Essentially, the preparation was intellectual and psychological, in the sense that there had to be a reasoned conviction of this position, and that there had also to be a readiness for the measure that would turn the conviction into fact. Coupled with this, the three countries prepared their public for the objective and the measures it necessitated, and for the confrontation it was expected to bring about. Elsewhere, whatever the conviction predominating, nationalisation measures and the take-over of control were to follow much later in the 1970s. Notwithstanding this differentiation, it remains true that, for the Arab oil exporters as a group, the 1970s were the deciding decade which witnessed the full fruition of the 'policy of control' after its gestation phase in the 1960s.

## How Was Control Achieved?

The discussion undertaken so far in this chapter has tried to answer two broad questions: Why it was important for the Arab oil-exporting countries to take over control of the oil industry? and Why had they failed to do so in the decades before the turn of the 1970s, except on a very modest scale and within very narrow limits? In the present section, we shall show how control was taken, that is, we shall identify the landmarks in the course of the process of take-over, and the modalities and instruments used.

As a preliminary step, it is necessary to explain that the 1970s witnessed, not merely the take-over of the power of policy-formulation and decision-making in the broad and basic sense of the acquisition of the legal authority that comes with majority equity participation or nationalisation, but in all other areas of decision-making as well, whether they flow from equity ownership and control or not. In other words, the sudden take-over involved the assumption of the power to take over certain decisions (such as price and volume determination) which the companies considered their own, even before equity control had taken place. The point is made here essentially to underline the unavoidable conviction that what brought about the transposition in the power positions of the two groups — the oil governments and the oil companies — was in the first place the change in attitude within the first group, or its new determination. It is true that the changed circumstances in the world oil market in favour of the governments, their expanded experience and increased self-confidence, together with the loosened and shaken network of the majors who had been placed under extreme pressure during the 1960s, were all supportive factors. But it is equally true that had the governments not decided to take the big step of control, the supportive factors would have remained only marginally strong in their impact on the powers and privileges of the two parties. There would have been modest quantitative improvements in the terms of the relationship, not a radical, qualitative transformation that drastically changed every aspect of this relationship.

The major landmarks of this truly revolutionary transformation fall into several areas, as Chapter 1 indicated. But we will only consider the policy of control here, that is, the take-over of ownership of the productive ventures which had been in the hands of the companies before, under the concessions or agreements which had governed their presence and activities. The remaining areas falling in the broad categories of upstream and downstream operations, and other policy areas relating to the utilisation of oil revenues and to regional complementarity, will occupy later chapters. Yet this division between policy areas must be understood as being rather artificial, designed as it is to make the discussion more easily manageable. The core of the matter is the determination of the producing countries to take the power of decision-making into their own hands; everything else follows from this basic position. Be that as it may, we will proceed with the division of the treatment as indicated earlier, concentrating presently on the policy of control by itself.

The point was made earlier that OPEC's main concern as an organisation was usually with price determination, and with other tax-related issues, as these determined government take in the final analysis. But OPEC has had other concerns, though never pursued as consistently and successfully before the 1970s. These included production prorationing (or programming, as it came to be called after the mid-1960s), conservation, and equity ownership. What issue was uppermost in the scale of priorities at the time depended on the industry's circumstances and the relative power of the two parties, the governments and companies. With prices under great strain in the 1960s, and the governments not yet sufficiently self-assured and their staffs inadequately trained and experienced, emphasis was mainly on prices and tax-related matters. But as the market tightened with the turn of the 1970s, it looked after prices and created the pressure for upward movement in them and in tax-related variables like the profit-sharing formula. In other words, the unilateral measures taken to adjust prices upwards in the early 1970s were market-led, as they were to be again by the end of the decade, particularly after the considerable drop in Iran's production as a result of the revolution which ended in the deposition of the Shah in 1979.

Substantially corrected and increased prices, expanded production, and increased revenues, made for a shift to two other concerns: equity control and conservation. It is developments in the first area that will now be traced. (Conservation will be discussed in Chapter 3.) As indicated earlier, the desire for equity participation preceded OPEC's foundation, but it did not take the form of effective pressure, although at times it constituted considerable 'intellectual' pressure, that is, among academics, technocrats, and certain schools of thought in the media. The resistance and resoluteness of the majors, with all the support they got from governments, defused the possible explosiveness of the pressure. The fact that certain independent newcomer companies had allowed equity sharing was discounted by the majors as a freak occurrence that could not be considered a precedent to be emulated under the 'most-favoured-nation' clauses in concession agreements. Iraq asked for participation in 1961; the refusal of the IPC was one of the major factors leading to Law 80 of December 1961 reclaiming all but 0.5 per cent of the concession area. Saudi Arabia itself, a long-term moderate in oil policy, hoped for participation, especially after the June 1967 Arab-Israeli war and the frustration and pressure for nationalisation of Western interests that the defeat of the Arabs produced. But there was no determined and immediately decisive follow-up there or anywhere else. It was generally felt that the conditions in the 1960s were not opportune for the confrontation with the companies that would have been inevitable.

The cry for nationalisation — outright full take-over of the ownership of the productive ventures in the hands of the oil companies — enjoyed much more appeal among those sectors of the population in the radical oil countries that were concerned about and involved in oil matters and national control over resources, as well as in government circles there. Elsewhere, the favourite mechanism for control was equity participation, that is, partial ownership of the ventures, rising rapidly to constitute majority control.

The debate between the advocates of nationalisation and those of participation was heated most of the time from the 1960s to the early 1970s. But, as hindsight now shows us, there was not much substance in the controversy, as the orientations and positions of the governments in the two camps translated into actual policies revealed in due course. Thus, none of the governments that had opted in theory for nationalisation did in practice nationalise all the companies operating on its soil in one blow; and none of the governments that had opted for participation failed in due course to acquire majority equity participation. Between the first day of the decade of the 1970s, when none of the Arab oil exporters had majority control, and the last day of the 1970s, all came to have majority control, and some came to have full control. The differentiated picture falls within these two not-too-distant limits, that is, 60 per cent and 100 per cent ownership and control.

Although the two courses of nationalisation and participation in practice converged in a matter of one decade, they were not identical. Behind each of them stood the political and ideological orientation of the political leadership, however defined, and the socio-economic system it espoused. These could be divided into two broad categories: the countries that were politically West-orientated, which believed in gradual change within an essentially capitalistic system based on a market economy and profit motivation, even if they allowed a large government sector to emerge and instituted a vast welfare system; and the countries that were either neutralist or East-orientated (essentially very critical of United States policy in the Arab region and the world), which believed in radical change within an avowed socialist framework, involving vast public ownership of productive assets as well as a welfare system.

As in the case of the convergence of the two courses of nationalisation and participation, the ideological/political and socio/economic orientations and systems, on analysis, do not add up to very different systems in application, and both now show strong signs of state capitalism, whatever else they claim to be.[22] None the less, it would be instructive to trace the two courses and see how each of the countries concerned with control arrived at its target by the end of the 1970s.

The earliest efforts to achieve control were individual efforts, even after OPEC had been established and had succeeded in forcing the companies, albeit after stiff resistance and devious efforts to escape, to deal with it collectively on behalf of its members. The latest attempt by Saudi Arabia to have Aramco (that is, the four American companies which constitute Aramco) accede to its demand for participation was in 1968. It failed in the face of the companies' obduracy, which continued to make them refuse to see the writing on the wall. This was particularly puzzling, since Saudi Arabia's position was moderate and allowed for continued extensive co-operation with Aramco. Four years later, the Saudi Minister of Petroleum and Mineral Resources was still willing to describe participation as 'a catholic marriage between the producers on the one hand and the consumers and the majors or independent oil companies on the other hand by linking both of them to an extent where it is almost impossible for any of them to divorce.'[23] Most wives — and we assume the companies to stand for the wife in this instance — would be certain to like such a marriage formula where their presence could not be dispensed with. Obviously, the companies valued their freedom of action, as being much more rewarding than even the security of the attractive marriage offered them.

OPEC duly was seized of the issue of participation, and in June 1968 issued a major document, entitled 'Declaratory Statement of Petroleum Policy in Member Countries',[24] which stipulated among other things that

> Where provision for Governmental participation in the ownership of the concession-holding company under any of the present petroleum contracts has not been made, the Government may acquire a reasonable participation, on the grounds of the principle of changing circumstances.
>
> If such provision has actually been made but avoided by the operators concerned, the rate provided for shall serve as a minimum basis for the participation to be acquired.

This was the most significant step taken by OPEC as a body in the 1960s towards control by its members of their oil resources and the activities and operations related to oil.

The companies were not slow to react, invoking 'the sanctity of contracts', as their practice had been on similar occasions when the governments had indicated the will and determination to act unilaterally. But, whatever the effect of the resistance, it was probably the governments' own slowness to go into confrontation at the time that was the major reason that participation only made headway some three years later. It is not unlikely that the individual preference of some governments for nationalisation was one reason why they did not press for equity participation. Two price agreements reached after considerable pressure, in Tehran and Tripoli in February and April 1971 respectively, had to intervene, before the governments turned to participation as a priority. It seems most likely that they wanted to dispose of the price question first; having done so and worked out a formula meant to last for five years, they felt they could tackle participation.

At this point, the governments faced the issue with new energy. In a conference in July 1971, OPEC resolved:

that Member Countries shall take immediate steps towards the effective implementation of the principle of Participation in the existing oil concessions. To this end, a Ministerial Committee shall be formed ... to draw up the bases for the implementation of effective participation ... and to submit its recommendations to an extraordinary meeting ... on 22nd September 1971. (Resolution XXIV.135.)

The sense of urgency was expressed in the setting of the date for an extraordinary meeting. This meeting was duly held and resolved:

(1), that all Member Countries concerned shall establish negotiations with the oil companies, either individually or in groups, with a view to achieving effective participation on the bases proposed by the said Ministerial Committee, and (2) that the results of the negotiations shall be submitted to the Conference for coordination. In case such negotiations fail to achieve their purpose, the Conference shall determine a procedure with a view to *enforcing and achieving the objectives of effective participation through concerted action.* (Resolution XXV.139. Emphasis added.)

The emphasised part of the Resolution is significant: it introduces enforcement; it talks of 'effective' participation (to warn against a semblance of participation involving only nominal equity acquisition); and it warns that concerted action would be taken if the companies balked, rather than weaker individual action. The Gulf group of OPEC members originally had in mind an initial 20 per cent participation, to rise later to 51 per cent. Libya, on the other hand, wanted an immediate 51 per cent participation. In its desire for majority control, it was guided by two factors: its own avant-garde position involving great concern for as quick a take-over of a determining position as possible, and Algeria's nationalisation of 51 per cent of the French companies operating in its territory in February 1971. The fact that several months had passed between Algeria's measure and OPEC's July meeting, was added evidence, if any was needed, of the mildness and reasonableness of the mainstream of OPEC membership, both as regards the timing of their Resolution, and the size of participation that they seem to have had in mind.

Once again, the companies balked and failed to see the writing on the wall, though it was getting increasingly legible. The grounds on which they resisted an agreement were outwardly the terms under which participation was to be effected; most probably, the real issue was the weakening of their control, and the terms came as a convenient alibi. The terms involved the cost of acquisition of equity, and the 'buy-back' price of oil that was to become the property of the governments.

We need not go into the details of the points at issue here, and the two positions in confrontation with respect to each of them. (The interested reader can refer to the specialised journals, as well as to Seymour's *OPEC: Instrument of Change*, for details.) It is sufficient to say at this point that negotiations were so tough and unproductive that the Monarch of Saudi Arabia had himself to issue a 'stern royal warning' (in Seymour's phrase), that unilateral action would be taken if the companies continued in their stand. This intervention was occasioned by Aramco's attempting last-minute diversionary action that in fact aimed at avoiding the question of participation in existing ventures and offered a larger ratio in ventures yet to be undertaken. Finally, the companies gave in, and accepted participation with an initial slice of 20 per cent. But this was only in March 1972. By that time, OPEC members were getting so impatient that they had almost come to the point of legislating participation at the level of 51 per cent and acquiring control in one stroke.

The negotiations for the terms of compensation and the determination of the buy-back price of government entitlement of crude took many months again. And once more OPEC had to warn that unilateral action would be taken if the companies continued with their style of using delaying tactics and offering unacceptable alternative terms, only hours before OPEC was to have a meeting for decision. While this was taking place, two very significant measures were taken by Libya and Iraq. The first nationalised British Petroleum assets because of Britain's failure to act to stop the occupation by Iran of three Arab islands in the lower part of the Arabian Gulf, which had been under British 'protection' shortly before. The second nationalised the Iraq Petroleum Company (except for the share of the Compagnie Française des Pétroles) on 1 June 1972, thus 'leapfrogging' the participation arrangements; however, the Mosul Petroleum Company and the Basrah Petroleum Company were not touched at the time.

A package deal was finally agreed in October 1972, involving gradual participation, compensation, and 'buy-back' of government crude entitlement under participation. The doses of participation were to be as follows: 25, 30, 35, 40, 45, and 51 per cent equity acquisition on the first day of 1973, 1978, 1979, 1980, 1981, and 1982 respectively. Iraq did not adhere to the arrangements insofar as MPC and BPC were concerned, preferring to go on with its nationalisation policy; while Libya saw that the pace was too slow, as it was bent on obtaining a minimum of 50 per cent participation right away (as it had done under a recent agreement with ENI, the Italian State Oil Company). The Kuwait government, while accepting the terms, failed to obtain ratification of the agreement by its National Assembly; the Assembly wanted a much faster pace of participation.

At this point, the dam of company resistance seems to have crumbled, with the result that majority control or full take-over accelerated within one year of the agreement. This agreement could have been seen as a force for stability, as it seemed to satisfy the aspiration of the governments, though not ideally. However, it turned out to be an incentive for more gains, sooner than provided for. The newly-discovered power of the governments must have made them wish to make up for lost time, during the long years when the companies had dragged their feet on even the mildest demand. The course of action during the following year was essentially characterised by individual action by each of the governments, but it ought to be remembered that the initial, major breakthrough had come about through collective action within the framework of OPEC.

The pace increased. In March 1973, Iraq and the IPC group (that is, IPC itself along with MPC and BPC) reached an overall settlement involving compensation for the nationalised northern fields and expansion in production and export capacity for Basrah fields. In Libya, between January and September 1973, most of the independents and some of the majors submitted to the governments' demand for 51 per cent equity participation, with varying degrees of grace and decorum. But the tempo of change increased further.

By 1974, not only had the terms of the agreement with the Gulf members of OPEC been bettered by Iraq and the Mediterranean oil countries, but the Gulf countries also wanted much larger participation. The Arab-Israeli war of October 1973, with all the dramatic and historical events and changes it brought with it in the fields of oil economics and power relationships, had intervened, and with it a severe tightening in the oil market. The circumstances had changed almost beyond recognition, and the power structure predominating before the war collapsed finally and irrevocably. Consequently, all the members of OPEC wanted majority control, and the companies gave in — this time immediately and with hardly any counter-manoeuvring.

Thus, Kuwait, Saudi Arabia, Qatar, and Abu Dhabi raised the level of their participation to 60 per cent, then Kuwait reached 100 per cent or full take-over in December 1975. Qatar followed suit with the Qatar Petroleum Company and Shell in 1976-7. Saudi Arabia finished negotiations with Aramco for 100 per cent take-over in 1977 (with retroactivity of terms to the first day of 1976), but the arrangements were not finalised, and no agreement had been announced by the end of December 1979 (or, indeed, by the time of writing early in 1982). Iraq completed nationalisation by 1975. The picture is mixed in Libya, with some companies retaining a 49 per cent interest, but most others were fully taken over. In Algeria all companies have been nationalised except for the CFP which holds a 49 per cent interest.

The radical change in the pattern of ownership in the 1970s has carried with it parallel change in the terms of new association with the erstwhile concessionary companies. These terms relate to the price at which the companies can now acquire oil, the privileges, if any, they have in priority treatment with respect to assured supply, and the price at which they buy equity oil from governments. They also relate to the margin of profit allowed per barrel of oil lifted. Finally, where companies have been entrusted with certain technical services (whether in exploration, production, or other fields), certain fees and/or privileges are stipulated for.

It remains to be added that, though the terms of compensation for equity acquired (or for nationalisation effected) were a matter of long debate and controversy, in the end the companies accepted what the governments had proposed in the first place, namely strict net book value of the assets acquired at the time of take-over. Even in cases where more liberal terms had initially been suggested, the final reckoning was on the net book value basis. The controversy in this context centered around the companies' claim for compensation for the profits foregone with regard to the oil reserves not yet extracted and sold. The governments categorically refused this demand on principle, since accepting it, no matter what the compensation turned out to be, would have amounted to an admission that the country's natural resources belonged to the companies. No sovereign state could accept such a claim, even if it were still to be under foreign domination.

## Instrumentalities of Control

Control was, and continues to be, effected through three instrumentalities and modalities, with one fourth instrument that has emerged in the 1970s. The first three are the individual oil governments which enjoy sovereignty, and in which ultimate authority resides to accept or refuse decisions and recommendations made elsewhere; OPEC as a collectivity consisting of seven Arab and six non-Arab oil-exporting governments, and whose authority and 'sovereignty' derive from the authority with which the governments endow it, both in its constitution and in subsequent resolutions; and OAPEC, which is the collectivity of ten Arab oil-exporting governments, whose authority and 'sovereignty' is again what its government members endow it with both in its constitution and in subsequent resolutions.

A paradox can be encountered on a first look at the structure and levels of competence and authority of the three groups. This is that, theoretically, OPEC and/or OAPEC acting collectively as a body could be expected to exercise a larger measure of control at any one time, than any of their members acting singly. This 'group power', as we have seen, was what constituted the secret of the strength of the group of oil majors in the early years of the 1970s and earlier. But the history of the oil industry in the Arab world, and of OPEC and OAPEC themselves, reveals that the 'policy of control' was implemented in its fullest meaning (that is, covering all the policy areas which were delineated in Chapter 1), only through the action of individual governments in the final analysis. Only in one policy area was action taken collectively, and successfully: pricing. But even here, the final prerogative was always in the hands of individual governments.

The explanation of the paradox is simple: it is that the oil governments have themselves defined the central area of their concern and collective action within OPEC to be pricing and price-related policies and issues, although Resolution 2 taken in the first meeting of the founders on 14 September 1960 in Baghdad, stated in addition, that 'The principal aim of the Organization shall be the unification of petroleum policies for the Member countries and the determination of the best means for safeguarding the interests of Member Countries individually and collectively.' Concern with other issues like

*The Policy of Control*

production prorationing or programming, participation, and resource conservation, was to occupy the centre of attention in later years, though it did not form part of the stated objectives.

On the other hand, OAPEC's purposes and objectives as stated in Article Two of its Founding Agreement are much more general and can include every type of policy area, thanks to the absence of specificity.[25] This article speaks of co-operation as being the 'principal objective of the Organization', and goes on to specify the avenues of action in pursuit of this co-operation. In particular, these are to:

(a) Take adequate measures for the co-ordination of the petroleum economic policies of its members.
(b) Take adequate measures for the harmonisation of the legal systems of the member countries to the extent necessary for the Organisation to carry out its activity.
(c) Assist members in exchanging information and expertise and provide training and employment opportunities for citizens of member countries in member countries where such possibilities exist.
(d) Promote co-operation among members in working out solutions to problems facing them in the petroleum industry.
(e) Utilise the members' resources and common potentialities in establishing joint projects in various phases of the petroleum industry such as may be undertaken by all members or those interested in such projects.

Division of labour between OPEC and OAPEC was defined and, in spite of the comprehensiveness of the objectives of OAPEC, this Organisation in practice avoids discussion of, and policy-formulation relating to, certain policy areas including pricing, volume of production determination, conservation, and participation. Many of the issues relating to these policy areas have come to receive considerable thinking, discussion, and research in OAPEC, both at the Secretariat and at seminars and conferences organised by it, but not at the Ministerial Meetings, as the annual reports and press releases of the Organisation testify. (The reports were prepared and presented by OAPEC's Secretary General only beginning with the year 1973, but have continued uninterrupted ever since.) This rather narrow restriction of the topics dealt with by OAPEC's Council of Ministers and its Executive Bureau is regrettable, since it would be useful for the Arab members of OPEC, in their capacity as members of OAPEC, to co-ordinate their stand with respect to the policy areas excluded from the field of competence of OAPEC within, rather than outside, the framework of OAPEC.

For the time being, it is very useful to put on record OAPEC's valuable contribution to the oil industry and to the Arab economy in general in the last field specified in Article Two of the Founding Agreement, namely, the establishment of joint projects in various phases of the petroleum industry. Indeed, the joint projects and institutions set up during the Organisation's lifetime of a little over one decade have come to translate in concrete terms some of the aspects of control and technical mastery of the oil industry, to which this chapter has been devoted. Four categories of the institutions and activities can be identified.[26]

(a) Research and studies at the Secretariat, ending in published material of high quality and critical value to the members and to the specialised readership. (Some of the publications appear regularly.)

(b) Seminars and conferences, ranging in size, but calling in leading experts from within and outside the Arab world, centering around various subjects and issues in the oil industry.

(c) Training programmes in the basics of the oil industry for government officials and those of national oil companies, including the establishment of a specialised training institute (The Arab Petroleum Training Institute).

(d) The setting up of a Judical Board for the settling of disagreements among OAPEC members.

(e) The establishment of five joint ventures in the fields of maritime transport, shipbuilding and ship repairs, petroleum investments, petroleum services, and engineering and consulting.

It can be seen even from this concise listing that the ventures, the Board and the Institute are contributive to the implementation of several policy areas. In addition, they promote co-operation and co-ordination of activities among the ten members of the Organisation, beyond the promotion of their individual interests. In brief, an examination of the record of OAPEC in its years of existence reveals that it has travelled a long way, from an institution undertaking rather elementary functions like arranging the administration and logistics of meetings, and the exchange of oil information, to one undertaking a wide array of important functions as we have attempted to outline. The transformation is particularly interesting to watch and useful to experience in view of the seemingly narrow focus and low sights the governments had initially set for it *in practice*, as against the ambitious objectives they had incorporated in the Founding Agreement. Often ambition is resorted to as an alibi to divert attention from the modesty of performance in effect desired. But the Organisation seems to have succeeded in escaping the fate that had been, on the evidence in hand, meant for it.

This survey and discussion of instrumentalities would remain grossly incomplete if we were to ignore a fourth instrument: the national oil companies (NOCs), under whatever name they have been launched. Every oil-exporting country has at least one such institution, and invariably they fall under the jurisdiction or supervision of the oil ministry (again, whatever the official name it has). These NOCs enjoy different degrees of autonomy and freedom of action, but they all share one important feature: fast-growing expertise, mainly in the marketing of oil, but also in some instances in exploration and the development of new oilfields, production, transportation, refining, and the launching of petrochemical projects. Parallel with the NOCs, there are other supportive institutions to deal with specific functions like training, drilling, and production operations (whether of oil, gas liquefaction, fertilisers, pipeline systems, or other aspects of the industry). Several of these institutions were established in the 1970s.

The urgency with which the NOCs and the other institutions have been set up is explainable by the fact that they had been seen as instruments essential to the translation of the policy of control into concrete fact. It is remarkable that the vast institutional build-up was almost all achieved in the decade of the 1970s (except for the NOCs themselves, most of which predated the 1970s). This necessitated crash programmes of conceptualisation, design, legislation, organisation, financing, and manning. The last task was probably the most demanding, and it called for intensive training and recruitment, both of nationals and 'aliens', the latter including a very large component of Arabs. (Algeria and Iraq are the best equipped in expert manpower.) Above all, the achievement is evidence that the oil governments had understood the policy of control to mean much more than mere equity participation or nationalisation, and had embarked on the satisfaction of the various conditions of control beyond ownership of assets.

Yet the speed with which the governments and the ministries of oil have equipped themselves in institutions and instruments, legislation, and manpower and skills could not but lead to a certain measure of experimentation and trial and error. This is evident in the rather large turnover in organisational forms and rotation of functions. While, admittedly, this is rather wasteful of resources, it is unavoidable. As indicated earlier, the Arab oil governments have had to go through their apprenticeship in policy formulation and the control of operations and management all in one decade, against stiff company resistance, world crises, and large consumer threats not only to the very existence of OPEC but to the physical control of oilfields itself.[27] On balance, the 'learning-through-doing' which the Arab oil governments have acquired is most valuable — certainly more valuable than a relaxed acquisition of technological and managerial skill handed over by the companies, but only over a much longer span of time. The challenge of defiance, crisis, and struggle is certainly most educative to a country or a group of countries with the will and the determination to pick the challenge.

A number of issues emerge as a result of the shift in the locus of policy and decision-making from the foreign concessionaire oil companies to the national governments of the oil-exporting countries, and the changing circumstances and priorities that the shift has brought with it. In turn, the issues have significant implications for the industry and the parties involved in it: exporting governments, companies, industrial importing countries, and developing importing countries. They also have implications for the upstream and downstream operations in the hands of the new masters and decision-makers, as for the manner and effects of the utilisation of oil revenues. The following chapters will examine the various policy areas, but an identification and a more probing examination of the implications of certain policy areas will be undertaken in the final chapter of the study.

If we were to group all the issues and implications in one formulation, we would say that the policy of control has resulted in providing the exporting countries with great opportunity, but has also imposed on them weighty responsibility. The opportunity and the responsibility stand in juxtaposition, and often in confrontation. How they can be reconciled is a central question that is well worth exploring, and the effort will be made to undertake such exploration.

# 3

## NEW POLICY OPTIONS IN AN INTEGRATED CONTEXT

The adoption and implementation by the Arab oil-exporting governments in the 1970s of the 'policy of control', or the take-over of the power of decision-making, opened the door to a number of other important policy areas which would not have been accessible but for the take-over. In none of these areas, except that of the pricing of crude oil in the export market, was a deliberate process of policy formulation and application collectively undertaken and pursued consistently. Nor, indeed, were integrated policies explicitly conceptualised and formulated even at the level of single countries, except in one or two cases. Yet, as indicated earlier on, behind measures taken stood policy directions or implicit policies in each of the countries concerned which, *ex post*, we can now discern as coherent policies. The movement of these policies as formulated by individual countries more or less in the same direction, or their convergence in due course, can be seen not as an accidental occurrence but as one submitting to the logic of the similarity of the broad interests of the countries concerned,[1] and the general soundness of the policies formulated. This convergence makes up for the absence of collective formulation and justifies the label 'Arab oil policies'.

The new policy options which the Arab governments found within their jurisdiction and field of action in the 1970s were far-reaching, and cover every area of action related to the oil industry closely or distantly. Taken all together, they can be grouped into policies concerned with physical aspects of the industry, including both upstream and downstream operations; with financial aspects, including pricing of crude oil and gas exports, the accrual of oil revenues and of 'surpluses' and the disposal of these surpluses, and the interaction between the physical and financial aspects and its implications; with the developmental implications of oil operations and revenues; with the integrational impact of the oil industry both within countries and within the Arab region as a whole; and finally with the role of oil in supplementing the region's efforts in its co-operation with other Third World regions to design and promote the New International Economic Order.

The present chapter will deal with the first two areas of concern listed, namely, upstream and downstream operations, and pricing, with its immediate effect on revenue accrual and surplus building. The discussion will have the objective of examining the policies involved and the issues they raise, clearing some of the misconceptions around these policies, and indicating the flaws and opportunities for possible correction of the policies. To the extent possible, the discussion and evaluation will attempt a blend of short-term and longer-term preferences of oil producers, without losing sight of the desiderata of oil consumers. Furthermore, some reference will be made, as the occasion calls for and justifies, to the historical background of the policies under examination; this is to say that the immediate past experience with regard to the locus of policy-making and the content of policies will be pointed to as and when necessary, for the true assessment of the significance of the tidal change that occurred in the 1970s.

It is necessary to end these introductory remarks with the reminder that the oil policies to be examined individually are closely interrelated and strongly interdependent. The exploration for oil, for instance, cannot be understood in isolation from the volume of production thought necessary and effected, which in turn is associated with the question of conservation, with the allocation of oil and gas between various uses as fuel and industrial feedstock, and above all with the price of oil and gas and the uses to be made of the revenue expected to accrue from the sale of hydrocarbon crudes and products. Yet it would be near-impossible to discuss all these questions together, which necessitates the division of the discussion into various policy areas, if it is to be manageable. This said, it is also necessary to say that an attempt will be made to identify and define the nature and impact of the various interrelations and interdependences between the different policy areas as the discussion proceeds. The attempt will help show how difficult it is for the producing governments to formulate policies that have far-reaching ramifications and complex relations with other policies within the oil sector itself, as well as implications for the whole national economies, for the Arab regional economy, and for the international community at large, with its developed and developing countries at the same time. To demonstrate the complexity will, it is hoped, reveal the difficulty of policy-making which is designed to get as much understanding and acceptance as possible, and will at the same time demonstrate the speed with which the Arab oil exporters have gone through much of their apprenticeship in policy-making in a fast-changing world, and often with less than good-will and sympathy from the industrial countries.

Finally, although this chapter sets out to deal with policies relating to upstream and downstream operations, it will not allocate equal space or emphasis to all the policy areas of relevance. One of these areas, that of the pricing of crude, will be dealt with briefly, although it is of major importance and occupies a central position in the whole array of policy areas. This is because it has already received the maximum amount of thinking and writing in the literature. Two other areas, the building of the infrastructure that the oil industry requires, and the distribution of refined products at the end of the downstream chain of operations, will only be briefly referred to − the first because it raises no special issues of far-reaching implications, and the second because there were in the 1970s only the barest rudiments of an Arab policy relating to it.

## Upstream Operations: Policies and Implications

The policy areas which fall under this heading are those of exploration and development of the extractive activity, production with conservation of resources as its counterpart, marketing, and the establishment and/or acquisition of the infrastructure needed for upstream operations such as gathering and separation networks, pipelines, loading terminals, and tankers. We need not go into fine definitions of any of these areas, since we start with the assumption of a fair degree of acquaintance with and interest in the oil industry which renders such definition unnecessary. Consequently, we move straight on to an examination of exploration policies and practices in the 1970s and the issues associated with them, against their background under the concessionary regime of the oil industry.

*Exploration*

Obviously, exploration is the first step in the process of the development of an oil industry. Normally, in the case of a country searching for oil, it would be absurd to pose the question: Why explore for oil? Nevertheless, in a situation like that with which we are dealing, namely, one in which certain countries are already established oil producers and exporters, the question is not superfluous, but very pertinent. This is because there are reasons to question the timing and the intensity of the search, and others to justify it.

The *prima facie* case for exploration and the subsequent development of oil wells that prove rewarding is to extend the volume of probable and proven reserves and to expand production capacity and facilities. But this justification is predicated on the argument, where it applies, that the reserves-to-production (R/P) ratio is low, and/or the rate of depletion of the resource is high, which necessitates raising the ratio and/or reducing the rate. Yet, it can equally forcefully be argued in certain situations that a country or a group of countries may feel more immune to pressures to expand production, if the R/P ratio is allowed to decline at a satisfactory pace from the country's point of view. While it is true that new reserves could be found through exploration and well development without that necessarily leading to the actual expansion of extraction, it is equally true that pressures for expansion would be difficult to resist in the latter case.

Such arguments and counter-arguments are not a mere exercise in debate; they are relevant to the case of the Arab oil-exporting countries. This is so because of the pattern of behaviour of the major energy consumers, namely the advanced industrial countries constituting the OECD membership. These countries have varied sources of energy including oil in some, and their total share of world primary energy is much larger than the share of the Arab oil exporters[2] (see Table 3.1). Furthermore, the OECD group is capable of undertaking very energetic and effective programmes of development of alternative (non-oil) sources whether conventional or non-conventional, while the Arab countries are handicapped in this respect by the inadequacy of scientific and technical knowledge and expertise, and by the scarcity of coal which is the most abundant source of energy in existence on a worldwide basis.

Yet, the OECD countries put considerable pressure on the Arab oil exporters (and the rest of OPEC's membership) to produce and export much more oil than the internal and regional needs of most of the Arab exporters call for. In other words, the Arab exporters are urged to attach a higher priority to (what is described as their) international responsibilities than to their self-interest. To put the matter bluntly, the Arabs are urged to provide 'bridging oil' or 'transition oil', that is, oil in enough volume to bridge the transition from oil as the major source of energy to the promotion of other conventional (and subsequently non-conventional) sources. Since the transition will be undertaken primarily by the industrial countries, both those with market economies and with centrally-planned economies, the leading oil exporters constituting the membership of OPEC will be shouldering an unduly large part of the responsibility of bridging. In other words, they would be required to deplete their valuable, nonrenewable resources at an unduly fast pace, for the sake and the convenience of the industrial world.

TABLE 3.1: **Proven Crude Oil Reserves in Arab Oil-exporting Countries and World, 1979 and 1980**

|  | Reserves (million barrels) | |
|---|---|---|
|  | 1979 | 1980 |
| Iraq | 31,000 | 30,000 |
| Kuwait | 68,530 | 67,930 |
| UAE | 29,411 | 30,410 |
| Qatar | 3,760 | 3,585 |
| Saudi Arabia | 166,480 | 168,030 |
| Libya | 23,500 | 23,000 |
| Algeria | 8,440 | 8,200 |
| Total | 331,121 | 331,155 |
| World total | 641,623 | 648,525 |
| Arab world per cent | 51.6 | 51.1 |

*Source:* OPEC, *Annual Statistical Bulletin 1980.*

In addition to the pressure on the Arab exporters to supply more oil than it is in their interest to do, or even in the interest of the international community (inasmuch as the easy availability of oil will reduce the urgency for the search for alternative sources of energy), many oil specialists and politicians in the industrial countries use a form of intellectual blackmail against the Arab exporters to push them into larger production than they want to undertake. The substance of the blackmail is that the development of alternative sources of energy, and unforeseen changes in energy technology in the coming years or decades, may well make oil much less vital to the world economy than it is now. If this happens, the argument goes, then the oil producers may regret any conservationist tendencies they have now. They would therefore be wise and self-interested to speed up production rather than restrict it on grounds of conservation of resources. This kind of approach obviously works in favour of accelerated exploration and development.

However, Arab producers are well aware of the frailty of this argument. They are confident that there will continue to be technologically advantageous, economically remunerative, and socially beneficial uses for oil and gas well into the foreseeable future no matter what other sources of energy come into prominence, generally as a vital input for industrialisation, and particularly in the fields of transport, fertiliser production, and the petrochemical industry. Furthermore, in thinking about the utilisation of hydrocarbons, the Arab producers have to take into sufficient consideration the fast growth in their own consumption of hydrocarbons and that of non-oil developing countries, Arab and non-Arab alike. While this factor is promotive of accelerated exploration and the establishment of greater reserves, it is equally promotive of great self-restraint in the face of pressure for expansion in exports to the industrial countries. It also calls for much greater discipline in energy consumption in the Arab and non-Arab developing countries.

According to estimates presented at the First Arab Energy Conference held in Abu Dhabi in March 1979,[3] total Arab energy consumption may well reach 10 million b/d of oil equivalent by the end of the century if not checked earlier; this volume would be equal to half total Arab oil production in 1980. Even if this estimate proved too high, the exportable surplus — other things being equal — would be much less than it had been in the late 1970s and the first year or two of the 1980s. In brief, the availability of bridging oil in the quantities desired by the industrial world (essentially the OECD community), would not be possible unless the Arab oil exporters (and other OPEC members) accepted expanding their production facilities and their production considerably, thus depleting their resources disadvantageously fast. If they were to do this, they would find themselves subsequently with excessively reduced reserves before having achieved a satisfactory level of development, while OECD countries would find themselves with considerable non-oil energy sources of the conventional, already available types, plus whatever new sources they may have developed by then, both renewable and non-renewable.

These concerns were expressed cogently in 1978 by Dr Ali A. Attiga, Secretary-General of the Organisation of Arab Petroleum Exporting Countries, in his opening statement at the First Seminar on Petroleum Exploration held by OAPEC. It is worth quoting at some length from his statement:

> In essence, the core of the main problems facing the oil-exporting countries, and especially the OAPEC member states, is the excessive world energy burden on their depletable oil and gas resources. This burden consists of having to supply the world's energy needs out of limited resources, while having to provide for the unlimited present and future needs of their underdeveloped economies. Thus, in the present-day energy transition, the oil-exporting countries are rapidly depleting their limited and most valuable natural resources in order to enable the oil-importing industrialized countries to maintain super-affluent societies. The ultimate objective of this energy transition is the attainment of energy self-sufficiency in the powerful industrialized countries, at which time the oil-exporting countries would have exhausted their hydrocarbon resources. It is obvious that if this happens without the attainment of viable economic growth and the development of alternative sources of energy in the oil-exporting countries, the future of these countries will be dismal indeed. I strongly believe that it is the responsibility of our generation to work together towards the prevention of such a future.
>
> One important and most immediate option in this regard is the increase in the life span of the oil resources of our member countries. This can be accomplished through a combination of increased supply and a more conservative rate of exploitation.[4]

The point was sharpened further by Dr Attiga more recently in his statement at the United Nations Conference on New and Renewable Sources of Energy, held in Nairobi, Kenya, in August 1981. Again we quote at great length:

> In this century the world has seen two major global energy transitions: the first was from coal and renewable sources of energy to oil and gas and the second is almost the reverse of the first transition. It is significant to note that the exploitation of the energy resources of developing countries played the key role in both transitions. Specifically, during and after the first and second world wars it was the exploitation of the oil resources of a group of developing countries in the Middle East, Latin America and Africa which accelerated the first major global energy transition. During that transition the developing countries, whether under colonial rule or after independence, did not benefit from the economic and technological advantages. Due to their almost stagnant economies their oil consumption was insignificant and they generally left the planning and management of their commercial energy requirements to foreign companies. This was the case for both the oil importing and the oil exporting developing countries. However, it was the energy resources of the oil exporting countries which carried the burden of the energy transition and it was the economies of the developed industrial countries which gained the most from it. But if that transition had continued it would have inevitably led the world into a fatal dead-end, because it was a transition from relatively abundant and renewable sources of energy to finite and easily depletable reserves of oil. The frightful result of that transition can be seen even today, eight years after the major oil price adjustment of 1973/4. Thus, although oil and gas represent only about 20 per cent of the entire world fossil fuel resources they still account for over 60 per cent of the world's consumption of primary

commercial fuels. It is certain that if the price of oil remained at the prevailing levels of the sixties and the early seventies the percentage of oil in global energy consumption would today be much higher while oil reserves as a percentage of global fossil fuels would be much lower, thus bringing the world to the brink of total economic collapse.

Even today we find that the reserve production ratio in the OAPEC area as a whole is only about 47 years compared with 96 years in 1960 despite the large oil discoveries during the 1960s. Largely as a result of the oil price adjustments in 1973/4 we are now witnessing a new kind of global energy transition. It is a transition from oil to other more abundant, but generally more costly sources of energy. But unfortunately, it is not that simple, because the developing countries require two kinds of energy transitions. First, they have to continue and even accelerate the transition from some renewable and non-commercial forms of energy to oil and gas in certain vital sectors of their economies. At the same time, they need to do all they can to develop their new and renewable energy resources in order to diversify their energy mix and prepare themselves for the post-oil era. This applies to all the developing countries, whether oil importers or exporters. In order to do that they have to depend, in the short and medium terms, on oil and gas, coal and nuclear power. This is why we, in OAPEC, strongly support the view that developing countries should have priority in the supply of all these sources of energy.[5]

The situation as we have described it presents the Arab oil exporters with a dilemma. They need to undertake active exploration and development of hydrocarbon resources in order to keep a satisfactory, or a gently-declining R/P ratio and to lighten the impact of depletion of resources; but they also need to conserve their resources to satisfy their own future expanded needs, and to protect themselves against the pressure of demand by the industrial countries, particularly against the attempt by these countries to treat Arab oil as bridging oil. The reconciliation between these two conflicting desiderata can be achieved through a policy like the one suggested in the first quotation given above, namely, 'a combination of increased supply and a more conservative rate of exploitation'.

Such a policy is pursued only in part now by the Arab oil exporters. There is keen awareness of the dictates of conservation in all oil-exporting countries, but the ceiling set on production in certain instances is too high by the admission of the country or countries setting the ceiling in question. (Nowhere has the ceiling been lowered to a level where it is critically close to the minimum called for by the country's need for associated gas in, for example, water desalination, electricity generation, and other basic activities.) Likewise, there is clear awareness of the imperativeness of speeding up the search for new reserves or for methods of improving recovery rates from existing wells, and the expansion of potential supplies, but this search is not pursued as energetically as the logic of the situation requires. Thus, the two components of the ideal solution that the Secretary-General of OAPEC suggests are both eroded in effect: through some neglect of conservation requirements, and through insufficient exploration activity. However, the policy of strict conservation and the issues it raises in practice will be examined as we come later to production policy with respect to which conservation constitutes the other side of the coin. For the present, we will dwell briefly on Arab exploration efforts under the policy of control, and the issues associated with them.

First, the facts. Without going into details, we can say that the exploration activity has on the whole been slack in the Arab region, compared both with exploration in earlier decades and with exploration during the 1970s in the United States and other industrial countries of the OECD. This is true whether we use as the criterion the volume of geophysical exploration (measured in party/month units), the number of wells drilled, or the number of rigs in use. Table 3.2 presents a comparison of data for the world (excluding the socialist countries) with respect to each of the three criteria, for the period 1970-9 divided into two sub-periods.[6] The table shows that the ten members of OAPEC have increased their exploration activity during the years under report, both in absolute and in relative terms. But, their overall position is much less favourable than that of the other regions. The very wide gap between the Arab region on the one hand, and North America and the other industrial countries on the other, needs some explanation, particularly in view of the much greater promise of large, remunerative finds in the Arab countries of the Middle East and North Africa. The disproportionately large number of giant oil fields found in the Arab region[7] suggests a strong likelihood of other remunerative finds, if not giant, then of medium and small size. The response to exploration, namely the average output per cubic foot or meter explored, remains the highest in the world, and is another reason to justify greater exploration activity in the Arab region.

The OAPEC study on exploration from which the table is reproduced reveals four points of great significance for exploration. These are: (a) most Arab oil and gas reserves lie in a small number of oil fields — indeed, more than 70 per cent of reserves lie in a very small number of super-giant fields (against 41 per cent in the United States), which suggests the great promise of yet-to-be explored medium and small-size fields; (b) 88 per cent of oil finds were discovered before 1966, leaving only 12 per cent for the period 1966-80; (c) the R/P ratio has remained stable at 46 years between 1973 and 1980, which suggests the need for greater exploration activity to compensate for future depletion and the declining rate of recovery owing to the ageing of oil wells; and (d) the response, in terms of potential output per drilling volume undertaken, is very high, particularly in the Middle Eastern part of the Arab region.[8]

These findings, plus the advantages which the Arab oil industry enjoys and which we mentioned in the preceding

TABLE 3.2: Comparison of Exploration Activity in Various Regions of the World During the 1970s

| Indicator | Geophysical exploration Party/month | | | | Exploration wells | | | | Yearly average of rigs | | | |
|---|---|---|---|---|---|---|---|---|---|---|---|---|
| | 1971-4 | | 1975-8 | | 1970-4 | | 1975-9 | | 1971-4 | | 1975-9 | |
| Region | No. | % | No. | % | No. | % | No. | % | No. | % | No. | % |
| OAPEC countries | 2,170 | 7.8 | 3,270 | 10.8 | 325 | 0.6 | 700 | 1.0 | 111 | 5.8 | 233 | 6.9 |
| North America | 13,840 | 50.0 | 17,430 | 57.6 | 47,100 | 86.6 | 62,650 | 89.5 | 1,212 | 63.2 | 2,331 | 69.3 |
| Other industrial countries[a] | 3,220 | 11.6 | 3,160 | 10.5 | 1,700 | 3.1 | 2,000 | 2.9 | 114 | 6.0 | 163 | 4.8 |
| Rest of the world[b] | 8,470 | 30.6 | 6,380 | 21.1 | 5,275 | 9.7 | 4,650 | 6.6 | 480 | 25.0 | 639 | 19.0 |
| Total | 27,700 | 100.0 | 30,240 | 100.0 | 54,400 | 100.0 | 70,000 | 100.0 | 1,917 | 100.0 | 3,366 | 100.0 |

Notes: a. Includes Western Europe, New Zealand, Australia, and Japan. b. Excludes socialist countries.
Sources: Paper prepared by OAPEC, Directorate of Exploration and Production, for the Second Arab Energy Conference (March 1982), Table 2, p. 29, entitled 'Oil and Gas Exploration in the Arab Homeland and its Future Prospects'. The table in the paper quotes the original sources of the information.

paragraph, strongly suggest the need for the intensification of exploration even if extraction were to be rationed in order to avoid excessive depletion of reserves.[9] Three issues will have to be faced if exploration is to be speeded up, within the context of a policy of expanded oil discovery and restrained production.

The first is the choice of the most productive formulas or modalities of activity whereby the desire for self-dependence by the oil countries is reconciled with the imperativeness of the recourse to foreign expertise and specialised equipment in almost all cases. No formula has been devised which is capable of making the erstwhile concessionary companies undertake much of the exploration. Indeed, their effort in this field had been slack even in the second half of the 1960s, particularly in those cases where they had rough relations with the national governments, but it has become minimal since 1973/4. However, this issue is easily surmountable in the medium and longer term. Within the medium term, the national governments can continue to engage the services of firms specialising in the various aspects of exploration work, precisely as they are doing now in most instances, directly or through the continued participation arrangements or other service contracts. In the longer term, the appropriate answer lies in the acquisition by the Arab Petroleum Services Company already established by OAPEC of the necessary expertise, facilities, and finance to undertake exploration work on a large scale for OAPEC member countries. The most serious bottleneck is Arab technical skills; all the other requirements can be readily satisfied once this bottleneck is widened.

However, a second issue will also have to be solved if the APSC is to function efficiently: this is close co-operation among the Arab oil producers in the area of exchange of information of relevance to exploration. A large body of data has been collected over the decades with regard to geological formations and past experience in the search for oil and gas. It would facilitate the work of each country considerably if the information collected and the experience gained were to be pooled and shared, with continuous updating and systematisation. The APSC will need the services of foreign experts for a number of years in highly-specialised activities using very modern techniques and equipment, particularly as the technologies involved experience continuous improvement. But this need can be easily overcome through the engagement of such services as are required, until Arab expertise has been acquired in adequate quality and quantity.

A third issue will be encountered in the dual attempt to expand exploration activity in the Arab region and simultaneously reduce the pressure on Arab producers to shoulder a large portion of production responsibility for the satisfaction of world demand. This is co-operation with non-oil Third World countries in their own search for hydrocarbon resources. The Arab producers have a stake in the success of such countries, inasmuch as the lightening of the pressure on Arab resources is of vital interest to the Arabs. The co-operation is essentially of a financial nature as far as the poorer countries are concerned. However, financial aid is not by any means the sole responsibility of the Arab oil countries; the advanced industrial countries and the World Bank have to shoulder a much larger portion of such aid. (One estimate puts the needs of these countries for exploration activities at $68.5 billion between 1976 and 1985 in 1977 dollars – well beyond the aggregate financing envisaged from all external sources.[10]) The industrial countries have generally delegated the responsibility for the determined search for oil to the international oil companies, under some formula acceptable to the countries and the companies concerned. On the other hand, the World Bank, which until recently shared the same attitude as the industrial countries,[11] has of late been converted to the idea of active participation in financing exploration in the less developed countries.

The design and implementation of formulas or modalities of co-operation between non-oil developing countries (both Arab and non-Arab), private companies, industrial countries, and the World Bank encounters significant institutional difficulties apart from financing problems. This has been highlighted by Francisco Parra, a noted international oil consultant and former Secretary-General of OPEC.[12] According to him, these difficulties include the desire of many governments of developing countries to keep the international companies out, because of a heritage of past fears of exploitation; the need for the companies to 're-orient their policies' in the direction of accelerating activity in the Third World along with the intensification of their activity in the industrial world; political upheaval and border disputes in many developing regions; the inadequacy of legal frameworks and inappropriateness of model exploration and production agreements to govern exploration activity; and the existence of certain restrictions on the availability of acreage for prospecting and exploration under some past arrangements with single companies or consortia of companies. The Arab oil exporters can be of help in smoothing most of these difficulties insofar as non-oil Arab countries are concerned. But their primary role for years to come will probably be as sources of financial aid. (Their expert help can only be forthcoming in the longer term, once they have improved their own capabilities.)

*Production/Conservation*

To put production of oil in juxtaposition to conservation is to underline the imperativeness of the consideration of the time span over which it is desired to extend the production of oil, out of the proved reserves at a given point in time. The imperativeness, quite obviously, arises from the finiteness of oil resources and their inevitable depletion. It acquires special force in the case of oil-exporting developing countries which have to insure against the eventual disappearance of their all-important oil resource. Such insurance can only be obtained by broadening the base of the economy and developing it as well as the society which runs it and for which it performs. Indeed, to be concerned about depletion at the peak of the power of the oil industry in the Arab region makes sense, like the taking by a young man of life insurance when he is still strong and active: in this case also the right time for an insurance policy is during the most productive years of one's life, not during the decline of energy and earning power when the premium of a new policy would be prohibitively high.

Yet, important as the pursuit of the proper insurance against depletion is, it will not detain us any longer here. For the moment, we will focus on production and its implications for the physical conservation of the oil resource. In so doing, we will also bear in mind that the policies and level of production, though formally determined by government decision, must also in large measure take account of demand for oil at home, in the Arab region, and in the world at large, as well as of the financial needs of the producing and exporting country as it defines those needs. Hence oil pricing and marketing policies are of close relevance to production policies and level. Though we are faced here with issues that seem to be almost inextricably tied together, we will fragment the discussion, leaving marketing and pricing out for the time being to be dealt with separately later on. However, the connections will be referred to whenever it is vital to do so for the cohesiveness and balance of the analysis.

In the decades before the 1970s, the main concern of the Arab oil exporters had been to sell as much oil as possible in order to earn a reasonably large volume of revenue. Hence, for example, the race of those Gulf producers, who were already relatively large exporters at the time, to take over fat slices of the production which Iran had been exporting when the nationalisation of the concessionary company in 1951 interrupted export from that country. The 1960s were also marked by substantial expansion in production (and export, since all production was then exported except for a tiny fraction kept for internal consumption). This is explainable by the emergence of Algeria (in 1961) and later Libya (in 1963) as relatively substantial producers, but much more particularly by the strong eagerness of the producers to maximise production in the face of low (even declining) prices, in order to obtain revenues capable of financing a respectable part of development programmes. Because of the fast growth just mentioned, the year 1970 stood at an index of 320.7 per cent as against the base year 1960, which represents an average compound rate of growth of 12.5 per cent per annum. The rate was barely lower for the half-decade 1965-70, when it stood at 12 per cent per annum.

Against this background of fast expansion in production, the 1970s present a much more modest picture. Thus the decade 1970-9 registered total net growth of 55.97 per cent, or an average annual rate of 5 per cent. However, the years 1970-3 which preceded the notable increase in prices had an annual rate of 11 per cent, as against a rate of 2.5 per cent for the period 1973-9.[13] If we compare production in 1979 with that of 1974, the first full year that sustained the impact of the rise in prices in 1973/4, we find that the net expansion during the years in question was 1.6 per cent, representing a negligible annual increase (see Table 3.3 for production data).[14] The substantial drop after the first few years of the 1970s is explainable by the impact on the importers of the rise in oil prices, and by the growing conservational tendencies among the exporters. Price increases had the double effect of inhibiting consumption and simultaneously making the expansion of production at pre-1973 rates unnecessary, thanks to the markedly increased return per barrel of exports, and the fast rise in total revenues.

The major policy line we can discern from this briefly drawn picture of production in the 1960s and 1970s is that the compelling objective of Arab exporters in the 1960s had been to expand production, with little or no concern for depletion and conservation, but that a 180-degree departure from this policy occurred in the 1970s, with major concern being for conservation and the slowing down of the expansion in production, if not reversing it altogether. The predominant policy of the 1970s was easy and to a considerable extent painless to adopt, inasmuch as the price increases permitted the exporters to make substantially increased revenues without a proportional increase in production — indeed, with a notably price-inelastic demand in the short run, revenues could rise along with a fall in production. This is a very good example of the possibility of having one's cake and eating it at the same time . . .

TABLE 3.3: **Arab Oil-exporting Countries' Crude Oil Production, 1979-80 (1,000 barrels per day)**

|  | 1979 | 1980 |
|---|---|---|
| Iraq | 3,476.9 | 2,646.4[a] |
| Kuwait | 2,500.3 | 1,663.7 |
| UAE | 1,830.7 | 1,701.9 |
| Qatar | 508.1 | 471.4 |
| Saudi Arabia | 9,532.6[b] | 9,900.5 |
| Libya | 2,091.7 | 1,830.0 |
| Algeria | 1,153.8 | 1,019.9 |
| Total | 21,094.1 | 19,233.8 |

*Notes*: a. Estimated. b. Revised.
*Source*: OPEC, *Annual Statistical Bulletin 1980*.

But it was not price developments alone (acting both on consumers and producers) that permitted and promoted the adoption of conservation. At least equally important was the growing awareness, which acquired its critical mass by the end of the 1970s and began to dominate thinking about reserves and production, of the finiteness and not-too-distant exhaustion of oil reserves, the inherent slowness of the process of development, and the need therefore to extend the life-time of oil reserves in order to enable the march of development to catch up with the extensive running down of reserves.

The dialectical relationship between the fast depletion of the 1960s, the slow unfolding of development, and the need to steer a safe course between slow movement and rush movement towards development – along with the political pressures associated with both rates of movement – was very forcefully analysed by the Minister of Oil of the State of Kuwait, Shaikh Ali Khalifa al Sabah, in a statement he made at the First Oxford Energy Seminar in September 1979.[15] His analysis led him to the conclusion, and policy suggestion, that a very high R/P ratio should be sought, and he opted for a ratio of 100 (that is, a one hundred years life-span for the reserves to last at the then-prevailing rate of production). Whether this ratio is unnecessarily high, or is reasonable, is beside the point; what is of great pertinence is the imperativeness of concern with R/P ratios appropriate for any given situation, taking into account the size of reserves and the volume of production, the country's size of population and its total requirements of oil revenue to meet its well-thought-out national, regional, and international objectives, and the R/P ratio base from which it starts. This last criterion is significant when the ratio is small, say, below 20 or 30, considering the slowness of the process of development which the Kuwaiti minister emphasised in his statement.

Designing an ideal blend of production and conservation desiderata in policy formulation is anything but easy. For one thing, no country can estimate in advance with any precision, how long it would take it to develop its economy to the point where non-oil income can compensate (in total or in large measure) for the loss of oil revenue owing to steep depletion of reserves. For another, no country can foresee the future trend of crude prices, or the prices of petroleum and refined and petrochemical products into which it may convert its crude oil and gas. Finally, nobody can predict what the future will bring by way of changes in sources of energy and energy technologies, or of changes in the sources and reserves of hydrocarbons themselves within the wide spectrum of energy sources. And the more distant the future being considered, the more difficult and less dependable the estimates, projections, or predictions.

Furthermore, another serious consideration influences the design of the blend of production and conservation to be built into policy decisions. This is the need to take into account not only national (and regional) consumption/investment/defence needs in terms of oil revenue (and therefore of oil production), but also the need of the international community for oil and the implications of the volume of supply available for world economic growth and balance of payments position. This component in policy-formulation has always been taken into consideration collectively by OPEC members in their pricing policies, and by these members formulating production policy individually. In fact, if anything, the Arab oil exporters have tended to attach excessive significance to international considerations and what they have come to accept as their international responsibility, often at the expense of their individual interests.[16] This is best evidenced by the excess of production over financial needs as assessed by the countries concerned in most instances, particularly by the ones with relatively smaller populations; the excess is translated into the build-up of financial reserves beyond all forms and channels of spending combined. These reserves, or surpluses, reached an estimated total of $178 billion by the end of 1979, and $275 billion by the end of 1980.[17] The size of these financial resources is an indication of the volume of production which the countries that own the resources could afford to do without. (Estimates of expendable production vary. Mr Adnan al-Janabi suggested half the 1979 production in a recent article.[18])

That production can be cut down without damaging most of the Arab producers (except, notably, Algeria and Iraq, which are high absorbers of funds owing to their diversified natural resources that are capable of being developed, and their relatively large populations), has not gone unnoticed by the producers concerned. Ministers and other persons in positions of responsibility have on a number of occasions indicated their awareness of the matter. But they justify their sustained production on the grounds that, to cut it sharply, would damage the world economy. Needless to say, a weakening of the world economy, particularly that of the United States, in which (or in whose currency) most of the surpluses are deposited or invested, would also harm the owners of the surpluses. But the counter-argument can be made that these oil exporters need not have had surpluses of the magnitude that they now have in the first place. Had they not interpreted their international responsibility as liberally as they have done, they would not have built up surpluses of the embarrassingly large size that we witness today.

The Arab oil exporters are impaled on one or the other of the horns of a painful dilemma. To continue with the current (1981) level of production (around 20 million b/d for the Arab OPEC members combined) in order not to dislocate the world economy, represents an unduly fast rate of depletion which allows the proven reserves to last for only 46 years at the rate of production of 1980 or 1981 – a much shorter time-span than that advocated by Minister Al-Sabah of Kuwait. On the other hand, to curtail production drastically and quickly, would disrupt the international economy and, in addition, would force the price of oil sharply upwards, thus causing added hardship, particularly to the oil-importing developing countries.[19] It would be difficult to find an instance where the reconciliation of self-interest with international responsibility would be more difficult. Probably the only possible and desirable reconciliation which would be least painful to the parties concerned would be one that moved gradually along three courses which converged on the same objective.

The three courses are: (a) more intensive and faster recourse to the use of non-oil sources of energy, with particular effort to be made by the industrial countries to develop non-oil alternatives and to promote an intensive search for new oil sources in developed and developing countries alike; (b) a fairer sharing of responsibility among all oil producers – not just OPEC members – to provide that part of energy consumption that hydrocarbons would be agreed upon to represent; and (c) to intensify conservation policies, practices, and technologies with respect to energy in general and oil in particular. This three-pronged approach to the problem, it ought to be added, can only be pursued over a number of years.[20] To attempt it in the short term would only create disruptive economic (and possibly political and social) stresses that it would not be possible to absorb.[21] The issue is involved and complex, as it touches on pricing, development, and aid policies apart from political questions.

We indicated earlier that the major concern of the oil exporters in the 1960s was the expansion of production in order to maximise revenues, given the stickiness of prices as determined by the oil companies which were then the decision-makers in this and other major areas. We further indicated that the concern shifted in the 1970s to conservation, thanks to the upward correction of oil prices and the growing anxiety over the fast depletion of oil reserves. These generalisations have to be somewhat qualified in two respects.

The first relates to the 1960s. Even in this decade, there were stirrings among OPEC members for some control over production, with a view to reducing the downward pressure on prices and avoiding the competition for larger slices of the export market by the producers. The overriding objective of this control, which was then called 'prorationing' was to ensure a fair sharing among OPEC members of the fast-growing market, particularly as the distribution of incremental exports after the stoppage of Iranian supplies in the early 1950s had been very uneven among Gulf exporters. The subject of prorationing was first broached within OPEC in January 1961, but nothing was done beyond preliminary discussion. It was reopened in 1965 (under the label 'production programming'), again inconclusively as there was a reluctance among most major producers to allow the subject of regulation of or control over production to pass from individual governments to collective competence.

Collective policy formulation at the level of OPEC first came in June 1968, when the 'Declaratory Statement of Petroleum Policy in Member Countries' was issued.[22] This Statement concerned itself with the 'exercise of permanent sovereignty over hydrocarbon resources' by OPEC governments, through a number of policy orientations. Owing to the importance of this resolution, we will outline its components or orientations. These included: (a) *development*, through the governments themselves exploring and developing hydrocarbon resources directly, as far as feasible; (b) *participation*, in respect of which 'reasonable participation' in the ownership of the concession-holding companies was urged; (c) *relinquishment*, in respect to which 'progressive and more accelerated relinquishment of acreage of present contract areas' was urged, including government participation in choosing the acreage to be relinquished; (d) *pricing*, to be based on posted or tax-reference price determined by government; (e) *fiscal stability and renegotiation*, involving an assurance to the operators of a fair return on their activity, with a warning that exorbitant returns to companies would not be tolerated and would lead to a renegotiation of terms between companies and governments; (f) *conservation* (and this is where concern lies at the moment), where the statement required the companies operating to observe and pursue the principle of conservation in order to safeguard the country's long-term interests, and enabled the government to lay down the necessary rules and procedures; and (g) *settlement of disputes*, where the Statement stipulated that such settlement was to fall within the 'jurisdiction of the competent national courts, or of specialized regional courts', not international arbitration, as the old concession agreements used to stipulate.

This formal Statement was not translated into concrete legislation or measures in the 1960s with respect to conservation (or development and participation, items (a) and (b), for that matter). Nor did conservation become a compelling objective of all OPEC member governments. Indeed, in June 1970 the Shah of Iran stated that his country would not be trying to get higher prices in order to increase oil revenues, but to achieve that end 'through more production of crude oil'.[23] Thus, it took the whole of the 1960s and the more-than-tripling of production, before concern over depletion and the need for conservation began to make itself felt strongly enough to lead to protective action — and even then not universally among OPEC membership. (Libya and Kuwait spearheaded precautionary action aimed at conservation, the former in 1970, and the latter in 1972.) The notable exception in the present context was Saudi Arabia, which did not go along with the mainstream concern over depletion. The vast abundance of its proven reserves, its special interpretation of the oil requirements of the international community and its probably exaggerated sense of international responsibility has made it continue to produce in quantities widely considered to be excessive. (Its R/P ratio stood at 46 in 1980, based on that year's proven reserves and level of production; however, in a recent speech, Minister of Oil and Mineral resources Shaikh Ahmad Zaki Yamani, strongly hinted that his country's reserves were exceedingly larger than those published, namely '173 billion barrels'.[24]

The second qualification called for with respect to the statement made earlier that the 1970s were marked by a keen awareness of the danger of depletion and the need for conservation policies and measures, is that the awareness may have been there among all OPEC members, but the reactions varied widely in determination and effectiveness. Many members set production ceilings, but these ceilings were not always the result of a careful consideration of conservation imperatives defined by the logic of the relevant variables. Furthermore, the ceilings for OPEC in the aggregate remained on the whole comfortably high for the consumers. Coupled with the fact that the price level between January 1974 and September 1979 did not rise in real terms – indeed went down[25] – the failure to impose strict conservation measures was serious. Its seriousness affected oil producers and consumers alike, the former because they did not capitalise on this conjuncture of circumstances to scale production down gradually for the sake of conservation and the avoidance of price stagnation or decline, and the latter because the comfortable supply situation and the declining prices weakened the pressure on them to search for alternative sources of energy and to be noticeably less wasteful in energy consumption.

The first serious look at prices and volume of production together, with a view to the design and maintenance of a rational relationship between them, came during an informal exchange of views undertaken by OPEC ministers in Saudi Arabia in May 1978. The exchange was to consider the elaboration of a long-term price and production strategy in preparation for the 1980s. A report on such a strategy was completed early in 1980 by a strategy committee formed for the purpose; it was accepted in February by six OPEC members and subsequently endorsed for discussion at an oil summit that was to be held in Baghdad in November 1980, on the occasion of OPEC's twentieth anniversary. The summit was not held owing to the Iraqi-Iranian war that broke out in September 1980, and the strategy has been shelved since then. The firming up of the market in 1979-80, and the considerable rise in prices as a result of the sizable drop in production because of the disruption caused by the war, have weakened the incentive for the examination and approval of a price-and-production strategy and related policies.*

Though the formulation of the strategy has remained an intellectual and professional exercise, and the strategy has not been translated into production/conservation/pricing policies, it has come as a confirmation and an expression of a slowly-building conviction of the need to move along from intensive concentration on pricing or on pricing-cum-production with the central objective being the maximisation of revenue for the producing country, to intensive concentration on volume of production with conservation as a central objective. Though the question of price remains, and will remain, very central in the concerns and deliberations of OPEC, it is now a matter of national and collective determination, no more one for negotiation with foreign oil companies.

Production policy itself is concerned with conservation, but has a wider focus than just conservation narrowly viewed. It is also concerned with the level of prices, inasmuch as this is a function of the volume of production; with the balancing of world demand requirements and supply policies; and with the developmental and other spending requirements of the

---

*The soft market of 1981 has weakened interest in the long-term strategy, and the first few months of 1982 witnessed a close dovetailing of prices and products.

producers. The shift in emphasis in the 1970s from prices to level of production has been spurred by the combination of three factors: keener awareness of the need for conservation, the exercise of full sovereignty and control over production during and after the war of October 1973 (involving production cutbacks and total embargo in a few instances by the Arab exporters), and the takeover of the industry either through majority or full participation or nationalisation.

Precisely as the oil exporters were prompted and helped in the 1970s by economic, managerial and political factors in their adoption of a conservationist outlook and policies, they had been inhibited in the 1960s by economic, managerial, and political factors from such adoption. These include the state of the market, the state of knowledge of oil trends and markets and ability to control them, the degree of control over the oil industry, and the interaction of these factors. But perhaps above all else the degree of self-confidence and the willingness to challenge the oil companies which had inhibited the producing countries in the 1960s, came to encourage them in the 1970s. The example and the urgings of Venezuela for the Arab exporters to control production, and the initiatives of Libya, Kuwait, and Iraq, were not sufficient in themselves to carry the whole Arab membership in the direction of production programming and control. Indeed, all along Saudi Arabia declared itself as having strong reasons not to apply restrictive production policies; and where the strength of argument supported restriction or programming, the lack of competence of OPEC to programme production was invoked, and shelter was found behind the sovereign right of governments to determine levels of production by themselves individually.[26] However, in fairness it ought to be added that in the late 1970s, Saudi Arabia became more amenable to the introduction of production levels as a subject for collective consideration by OPEC, although that did not lead to actual collective policy formulation of relevance.

The discrepancy between the outlook of Venezuela on the one hand, and the Arab exporters on the other, where the former became conservationist in the later 1950s while the latter were converted to conservationism only in the 1970s, is attributed by Seymour to the contrast in resource endowment. Venezuela had modest reserves, and it was wise for her to deplete them slowly; the Arab exporters on the whole had much more substantial reserves and in many instances had a later start, which led to a postponement of their concern with conservation. Furthermore, the Arab exporters were adding considerably more to their proven reserves than production was depleting those reserves, except in the 1970s. Drawing on Seymour again, while additions to reserves in the 1950s were over 12 times aggregate production, and were still about 6 times in the 1960s, they kept rising in the 1970s and only declined to the ratio of 50:65.[27]

But important as the foregoing explanation of discrepancy is, it has probably to be supplemented by some additional considerations, particularly if we are to consider policy imperatives for the future. One of these considerations is the fact that the Arab exporters have to depend excessively heavily on the oil resource, given the low level of development of their non-oil economies. This makes it necessary for them to deplete their oil resources at a carefully and delicately measured pace, in line with their ability to absorb investments for development, to diversify their economies, and to build up their internal productive capability, human, physical, and institutional.[28] To continue with production at levels higher than the pace of development (and vital consumption) requires, would be wasteful and dangerous, especially if it led to the piling up of huge financial reserves abroad. It is common knowledge now that these reserves are being eroded in value at a fast rate, owing to inflation and the fluctuation of the dollar in which much of the reserves is stored. Furthermore, the high average per capita income in the oil-exporting countries is to an important extent an illusion of serious implications for the future, if we consider the fast depletion of oil with which it is associated, the difficulty of making sound, production investments abroad (beyond the mere financial placement of funds or their deposit in banks), the deterioration of the work ethic in the financially-rich exporting countries owing to the weak linkage between effort and reward, the dangerous confusion of monetary riches with development, and the permissiveness that characterises consumption outlays and the determination of investment priorities and outlays.[29]

There is yet one other serious aspect that must be associated with unduly large production and the need for conservation. This is the high rate of increase in energy consumption in the oil countries beyond what is warranted by the requirements of productive activities, transport, and a reasonable rate of growth of household consumption. As mentioned earlier, if the growth is not checked soon, consumption by the year 2000 would divert to internal use in the Arab region about one half of the current production of 1980 or 1981, according to its present pattern and rate of increase. Such a development would leave for export to non-Arab markets only about half of what is currently exported.[30] This would have serious physical and financial implications for the oil exporters and importers, particularly if by the turn of the century alternative energy in sufficient versatility and quantity, and at tolerable prices, has not been found or developed.

The permissiveness in consumption in the oil-exporting countries, and also (though to a lesser extent) in the other oil-producing countries, speeds up the depletion of the resource by a sizable factor. And the seriousness of this over-consumption extends well beyond present wastefulness and permissiveness into the future, inasmuch as consumption levels, if allowed to harden through faulty habits, traditions and policies, will be most difficult to correct downwards; indeed, they could then only move upwards, particularly as a result of increases in population, greater urbanisation, the pressure of imitation and demonstration, and the much wider use of electrical and gas-using appliances and of private transport.[31] A substantial part of the pressure for high consumption is economic and social, and to a considerable extent it arises from paternalistic policies which allow the prices of oil and gas to be artificially very low. The need for the correction of these policies has only recently begun to be felt, and so the 1970s did not witness the formulation of conservationist policies addressed to domestic consumption. In our estimate, it will be some time before such policies will find their way to formulation and implementation. But they would not be sufficient in themselves for a satisfactory correction of consumption patterns and levels, unless supplemented by energetic educational campaigns aimed at discipline in consumption, and by the search for and use of energy-saving technologies and techniques. Furthermore, productive activity which is sheltered now from a realistic pricing of energy and thus receives a thinly-disguised subsidy, ought to be gradually exposed to as nearly full a costing of energy as possible, if the entry of its products into the international market is to be an honest indication of price and quality competitiveness.

Thus we see that production policies, and such closely-related matters as exploration, reserves, depletion and conservation, consumption, and of course pricing, raise grave issues of concern to oil exporters and importers alike.[32] These issues seem on the surface to be in contradiction, depending on whether they fall within the area of concern of exporters or importers. But closer examination will reveal that if foresight is applied, and true interdependence is sought, much of the concern for several of the issues will be shared, inasmuch as they relate to deeper, long-term common interests of oil exporters and importers alike. Without listing the main concerns of the two groups at this point, we will single out what seems to us to be the one major concern of each, in order to show that basically they have a common fundamental concern.[33]

The oil exporters are eager to produce and sell a volume of oil which, on the one hand, will bring in a volume of revenues capable of meeting their basic financial requirements for development and security, as well as for regional and international responsibilities which they wish to shoulder. On the other hand, they would like to conserve their oil resources against fast depletion, in view of the facts that the development process is very lengthy by nature and very demanding in financial, physical, and human resources, and that they have a strong preference for the possession of a national source of energy for their own future use. It is desirable for the exporters, having opted for a certain volume of production, to feel immune from severe fluctuations of this volume and of the revenues it brings in. Likewise, given the volume of production, they would like to increase the value added to it inside their economies through refining and industrialisation, thus tightening the integration of the oil sector with the rest of the economy.

The major oil importers, on their side, would like to have security of supplies[34] and gentle changes in current prices (if not outright freezing), but would like the volume of supplies to be a residue to be determined after the various sources of their domestic energy supply have been tapped. (The preference structure is more complicated, in that it contains a differentiation between cheaper and more expensive sources, and conventional and non-conventional sources. The sources tapped are those already developed or being developed, while alternative sources tend to be slowly explored and developed so long as cheaper oil is available.) Furthermore, the major importers would like to have as large a portion as possible of the revenues received by the exporters returned (re-cycled) to them for goods and services purchased.

Fundamentally, both parties are interested in the emergence of significant alternative sources of energy. Even as far as the long-term interests of the importers are concerned, the volume of oil exports must not be left as a residue to be determined by the availability of other sources of energy, but should enjoy predictability.[35] And predictability suits the interests of the exporters as well. Furthermore, a volume which lengthens the life span of oil reserves is in the true interest of both parties. How long a life span is a question which need not be and indeed is not beyond the ingenuity of competent, well-meaning, and responsible people to answer. Once the mutuality of interest has been accepted, the level of prices assigned by the exporters will be seen as reasonable in order to motivate the industrial countries, which are the major importers, in the search for credible alternative sources of energy.

It is obvious from this brief reference to the major issues facing exporters and importers that the problem for both groups of coming up with a blend of self-interest with international responsibility acceptable to both (or at least less unacceptable than any other blend) is not an impossible task. But it is a task that cannot be left to the operation of market forces alone, given the complexity of the several issues involved. The market by itself is not capable of bringing about an optimal situation from the point of view of the producer or the consumer. (The latter tends to ask the producer not to intervene, but to leave market forces to work on their own. But this position is usually taken only when it is believed that market forces will serve the interests of the consumer.) Hence the intervention of governments with corrective policies at the exporting and the importing end alike.[36] This is because there are social costs and returns, as well as political desiderata, that have to be considered, along with purely economic costs and returns.

All we can say in closing this section is that the Arab oil exporters, in spite of their awareness of the need for production/conservation policies that serve the long-term interests of their hydrocarbon resource, and of their economies and societies at large, have until now not resorted to such restrictive policies as are suggested by their self-interest, but have instead shown a large measure of concern for the interests of the international community as well. This is true with respect to the advanced industrial countries to whom oil supplies flow in a volume that would have to be curtailed if pure Arab self-interest were to be obeyed, and it is true with respect to the developing oil-importing countries to whom Arab aid has been flowing in substantial volume, to help finance oil imports and development in general. If anything, the Arabs have paid unduly large attention to international responsibility at the expense of the interest of the advanced countries themselves, in that stricter pricing and conservation policies would have led these countries to pursue a more energetic

## Marketing

As only about 7 per cent of the oil produced is consumed in OPEC member countries, exporting oil to foreign markets acquires special importance. Prior to the takeover of control and operations by the Arab governments in the 1970s, the marketing function was performed by the oil companies. And initially, before the entry of the independents into the industry, these companies as we have seen were the seven (or eight) majors, which handled marketing as well as all other operations called for by the industry within an integrated system. As Dr Fadhil Al-Chalabi describes clearly in a recent work, the system was integrated horizontally, in the geographical sense of the term whereby the companies' production and sale of oil complemented each other within the same phase of the industry; and vertically, in the sense that operations at one phase 'fed' the next phase.[37] The allocation of the oil produced by the majors was thus performed in both senses of integration: horizontally, whereby supplies went to partners in the consortia or in the oil cartel and to subsidiaries, and vertically, whereby the receivers of the crude used it as an input in their refining, and subsequently their distribution activities.

The entry of the independents into the industry changed this pattern of relations and movement only partially, as most of the operations of production, marketing, and pricing, as well as refining and distribution, continued to be undertaken by the majors, which controlled most of the oilfields, the pipeline systems and loading terminals, the tankers or tanker contracts, the refineries, and the distribution facilities. But, it must be noted, the breadth of their ownership and control narrowed somewhat as one moved further away from the oilfields.

The title of this section of the chapter is in fact somewhat misleading as there used to be no real oil market in the strict sense of the term under the regime of the cartel.[38] The majors did not sell crude oil to third parties. As crude oil is not used directly but in the form of refined products, it went to the owners of refineries – that is, to the oil producers themselves, their partners in the various consortia, or their subsidiaries. Thus, the companies produced for their own needs, or those of their associates which were deficit companies: that is, ones whose production fell short of their refining and distribution needs. Under the system, aggregate supply and demand were continuously in balance except for marginal, short-term aberrations arising from transient factors. The balance was achieved by design, not through the operation of an 'Invisible Hand' shrouded in mystery.

Arab advocates of nationalisation in the decades preceding the 1970s found themselves faced with the prospect of the inability of governments to market the oil except to the majors, which could not be expected to be accommodating under forced nationalisation. Invariably, the experience of Iran after nationalisation in the early 1950s was invoked to frighten off the advocates of nationalisation. With hindsight, one may wonder if marketing would have been as unmanageable as it used to be portrayed, if two or more acted in unison for a long enough period of time. However, as this condition would have been extremely difficult to satisfy, and the national states involved did not in those days enjoy full control and sovereignty in the true sense of the terms, their caution and shying away are understandable. They were kept away also from all other upstream operations, as these were meant to lead ultimately to the export of oil; since such export opportunities were only open for the majors, the governments were kept away from investments in all earlier phases. The scarcity of funds and of national manpower trained in the necessary skills further confirmed the full control of the majors.

This control was only gradually eroded, and the closed market circuit was gradually freed. It is now established in the recent history of the industry that these developments came from the side of the major consuming countries themselves, particularly the United States. As we suggested in Chapter 2, there is more than a touch of irony in the fact that United States policies ended by bringing about the erosion of the power of the majors and the take-over by the producing governments of power and control over the industry. This shift was the result of the building of many new refineries at the consumers' end instead of the producers' end, and the resultant opening of new marketing outlets other than those exclusively owned by the majors; the emergence (again with encouragement) of smaller, independent oil companies that sought all free outlets available; the emergence of national oil companies, NOCs, in the producing countries, for whom the outlook brightened as the widened outlets made themselves felt; and, in the early 1970s, the liberalisation by the United States of the import of oil owing to the pressure of consumption on domestic supplies, at the same time as supplies from the Western Hemisphere began to reflect strongly the impact of depletion. As the crack in the 'closed circuit' of the fully-integrated system of the majors widened, their solidarity weakened, even if marginally, and some competition among them began to be experienced.

OPEC national oil companies had their real opportunity to become a factor to contend with in the area of marketing, even if their transactions were small quantitatively, thanks to participation or nationalisation measures taken as of 1967 (in Algeria) and 1972 (in Iraq). Subsequent participation agreements between the governments and the companies, and nationalisations, put at the disposal of the NOCs more oil to market. By the mid and late 1970s, the exclusivity of the market had been overcome at least as far as the legal and institutional factors and frameworks were concerned. Thus, theoretically at least, the producing governments had the power to control the marketing of an increasing proportion of their production. This proportion had reached some 70 per cent of total production by the end of the decade, taking into account nationalisations, and full and partial (60 per cent) participations − the modalities that apply to the seven Arab members of OPEC in their relations with the foreign companies. But in practice, the producing governments, and/or the NOCs acting on their behalf, handle directly perhaps less than 25 per cent of the exports of the countries concerned as a group; some estimates put the proportion much lower at 10 per cent only. The rest, somewhere between 75 and 90 per cent of exports, is lifted by the oil companies (mainly the majors which had been or still are in operation).[39] Liftings by companies include straight sales by governments or NOCs, participation oil (the share of the companies concerned where they are still equity owners), and buy-back oil (the share of governments within participation arrangements, sold to the companies to deal with).

The detailed marketing arrangements between the governments or NOCs and the foreign companies need not detain us here, as we are primarily interested in the policies and issues associated with marketing. Nor do we need to trace the course which the margins of the majors have taken on every barrel of oil lifted during the 1970s. It is sufficient to make a few pertinent general remarks. The first is that the bulk of sales is still effected through foreign oil companies, but mainly through the majors; in other words, direct sales from producer to consumer government still constitute a much smaller proportion than indirect sales through an intermediary. Secondly, such three-party transactions are mainly made under special arrangements involving an agreed margin (a discount on the announced selling price), where partial participation agreements are still in force. Thirdly, the role of the majors in a pattern of participatory arrangements is, understandably, much larger than that of independent companies. Fourthly, margins or discounts are allowed even in some cases where nationalisation or full participation has taken place, and sales are made under long-term contracts.

The margin or discount allowed in buy-back and other contract liftings started by being quite substantial when participation arrangements were first put into effect, especially during the notable rise in prices in the first half of the decade. But the end of the decade saw the margin dwindle to something ranging between 15 and 70 cents per barrel, depending on the level of participation in the individual country and the attitude of the government to the oil companies concerned. Understandably, the size of the margin was a cause for concern to the oil countries during the early years of the 1970s, when it was a matter of dollars per barrel, rather than cents, but it became satisfactory after its adjustment.

One aspect of the concern was the disadvantage which the margin constituted for the NOCs which had to sell at the full price set by their governments to other parties which did not enjoy a reduction, such as consuming governments and oil companies outside participation arrangements. But the disadvantage has become smaller as the volume of sales by NOCs has increased somewhat and as the size of the margin has got much more modest. The latter development came about thanks to a substantial increase in the tax rate and in the royalty paid by companies in participation arrangements, as well as by a direct cut in the margin allowed. The expansion in the volume of sales by NOCs is such that the majors are now merely self-sufficient in oil, in the sense that they merely get enough oil for their refining and distribution requirements via their participation shares, or are even in deficit, needing at times to have to buy a certain small part of their downstream processing needs from OPEC countries.[40]

If the issue of the margins obtained by the oil companies and the comparative disadvantage suffered by the NOCs as a result is largely solved now, other issues linger on and do not seem to be in the process of being solved. Four of these will detain us here. The first is the still very limited marketing role of the NOCs owing to the continued existence of an integrated system of oil availability, transport, refining, and product distribution in the possession of the majors, despite the shrinkage of the dimensions of this system and the disappearance of the power of the majors in price and volume-of-production determination. The replacement of partial by full participation in several Arab OPEC countries will go some distance towards the expansion of the marketing role of the NOCs — but only theoretically. What is at issue is the expanded capability of the NOCs to supply the refineries of the industrial countries with their oil requirements directly in preference to the major and independent oil companies that lift these requirements themselves, and, ultimately, the ownership by the NOCs (or other agencies of the producing countries) of some refineries and distribution facilities in the large consuming countries, as well as the expansion of refining capacity in the producing countries themselves, so as to make possible the export of a considerable volume of refined products. Furthermore, as indicated in Chapter 2, the NOCs have yet to co-ordinate their activities within some agreed general framework that encompasses upstream as well as downstream activities. So far, little effective co-ordination on any question has taken place, let alone substantive questions.

The second issue to raise is of a much more general nature. It is the striking asymmetry between the major oil companies and the oil exporting countries with respect to the co-ordination of policies and the integration of activities, except in the area of pricing (and even there, with obvious difficulty). What is at issue is not merely, or even mainly, the weakness of the resolve to ensure such co-ordination and integration, which is a subjective factor, but the objective difference in situation. In this latter respect it must be noted that the oil countries lack the pattern of horizontal (geographical) integration which evens out the supply and flow of oil with the specific needs of particular markets, as well as the pattern of vertical integration (between upstream and downstream phases), which had the effect of making the oil industry in the hands of the majors one whole integrated operation into which supply and demand volumes fitted comfortably and harmoniously with no major maladjustments. And the likelihood is very slim, if not altogether non-existent, that the oil countries would achieve the same degree of integration as the major companies enjoyed until the late 1960s.

Two main reasons explain this assertion. These are the fact that the oil exporting countries are sovereign states that guard their sovereignty so jealously as to refuse to surrender enough of it to make close integration possible; and the increasing importance of non-OPEC oil suppliers which naturally reduces the degree of control that OPEC can have on the oil market, in a world where oil as a source of energy is losing some relative importance among energy sources in general.

The third issue to consider with respect to marketing is the absence of production programming within OPEC as a group, and therefore the exposure of OPEC countries to sharp ups and downs in the course of supply, particularly in a demand situation influenced considerably by conservationist tendencies among the large consumers, strong substitution pressures between oil and other sources of energy, and sluggish rates of economic growth. The interplay of price and demand factors (with the substitution factor implicit in both) is bound to lead to alternating excesses and shortages in supply. The effect of such relatively sharp fluctuations is in turn bound to influence prices sharply, where all the variables mentioned fluctuate considerably. Production programming, and therefore export predictability, is capable at least of stabilising one of the variables, and therefore reducing the scope for sharp fluctuations in the other variables, to the benefit of producers and consumers alike — provided that the programming is realistic and not arbitrarily and excessively restrictive.

The fourth and last issue to consider is the delicate balance needed between allowing margins or price discounts to foreign oil companies lifting Arab oil that are in possession of the necessary financial and technical capabilities or connections to undertake active exploration programmes, on the one hand, and containing the size of the margins or discounts so that they may not seriously inhibit the marketing activities of the NOCs. The linkage between marketing via foreign oil companies and exploration, which is necessary at the present moment and will continue to be so for some years, must not however be looked at as a long-term necessity. Ultimately, individual oil countries, or, preferably, OAPEC members collectively, ought to set their sights on acquiring the capability themselves to explore for and develop oilfields.

Though such acquisition is not easy, it is none the less quite possible within a reasonable span of time, given adequate financial resources, and assuming the necessary determination and the formulation of the appropriate policies and measures. Many steps have been taken in this direction, by OAPEC collectively and by individual Arab countries singly. But much more has still to be undertaken. Kuwait's recent (1981) purchase of the well-known company, Santa Fe, which is renowned for its far-reaching activity in exploration and its outstanding expertise and facilities, is a step in the right direction, provided of course that Kuwaiti and other Arab personnel are given the opportunity to participate actively and increasingly in the work of the company, and that the company may ultimately become effectively arabised. Arabisation which is optimal in the present context involves much more than the control by Arab capital of equity and the heavy participation of Arab manpower in the labour force of the company: it involves the true acquisition of skills by the Arab component of this force at all levels of skill and responsibility. If the acquisition of Santa Fe is one imaginative and bold step, other steps must also be taken, depending on the various needs and endowments of the different countries and of OAPEC as an organisation.

*Pricing*

The pricing of gas is excluded from the discussion in this section, partly because so far this has not been established on nearly as firm and clear a basis as that of crude, and partly because gas exports trail far behind those of crude oil in value. While the exports of liquefied gas are significant only for Algeria among the countries covered in this study, those of crude oil are significant for all seven countries.

Oil pricing occupies a central position among oil policies, because it is of direct and strong relevance to production and revenues, which are in turn of strong relevance to reserves/ depletion/conservation, the search for oil and other sources of energy, and also to the development of non-oil sectors at the national and regional levels, as well as to aid within the region and in the Third World at large. Yet, significant as oil prices are, they are not determined in and by themselves, but via the prices which petroleum products fetch in the market. Thus, the price of oil is a derived price. In other words, the price of crude is worked backwards, as a residue after the various costs (transportation, refining, distribution) involved in the transformation of crude into the 'basket of refined products' have been deducted from the pre-tax prices paid by the consumers for products.

Yet, having said this with reference to the prices of refined products being determined by market forces, it is necessary to point out that market forces do not determine crude oil prices all by themselves, in spite of the preponderant power of these forces. In fact, though to a much lesser extent, the price of crude is also the result of the operation of institutional forces: the decisions of individual OPEC governments or of OPEC acting collectively. The attribution of major influence to market forces is becoming truer as oil's share among all energy sources is getting smaller, and OPEC oil itself is losing some ground to non-OPEC oil. The two factors combined reduce the strength of institutional influences. In fact, it would be fair to say that oil prices have become more exposed to the operation of market forces since 1973 and the take-over of control by OPEC governments, than they had been under the regime of the cartel of concessionary companies. This is only in part the result of the operation of the two factors. One must remember that under the company regime, prices were set artificially low since they were merely an internal accounting mechanism within the closely-integrated system of the companies.

This last contention cannot be seriously challenged. Pricing before the onset of the 1970s failed to take into account movements of the general price level in the world, particularly in the industrial countries which are the major consumers of Arab and non-Arab oil. The companies did not allow for the correction of oil prices to make up for the three or four decades preceding the 1970s when these prices remained isolated from market forces. The isolation was motivated partly by the requirements of the integrated system of the oil industry, and partly by the concern of this industry for the political and economic interests of the Western governments to which the companies belonged. Consequently, the steep rise in prices effected by the OPEC governments beginning with the autumn of 1973 was not an arbitrary measure undertaken merely as an expression of sovereignty after the take-over of control, in violation of market forces, but one essentially undertaken to take into account the correction that had been grossly overdue, and the inflationary pressures generated in the industrial countries. In other words, the correction indicated sensitivity to genuine, undisguised market forces.

At this point it will be both useful and necessary to present a number of pertinent comments on pricing before 1973 in order to establish the case for the need for correction which was initiated in that year. These comments were made recently by the President of the Arab Fund for Economic and Social Development, Dr Mohammad Imadi.[41] According to him, that pricing formula

- reflected only the control of the producing oil companies, not the wishes of the states that owned the oil resources;
- did not observe the principle of marginal costs in the more complicated or difficult oilfields in the Arab countries; in addition, the costs accounted for were lower than the world average;
- was based on production costs below those encountered in the production of alternative sources of energy, a factor which led to a drop in the share of alternative sources and a rise in the share of oil in total energy consumption, therefore in the acceleration of depletion of oil;
- did not set oil prices at a level that would encourage the development of new alternatives to replace oil which is a depleting resource;
- incorporated government take (return to government)

which was very low; this take did not exceed 17 cents per barrel in the early phases of exploitation, or 90 cents by the end of the 1960s;
- did not take into account that oil was a depleting commodity, unlike renewable agricultural or industrial production; and
- permitted a decline in oil prices at a time when the general price level was rising. Thus, during the period 1960-9 the price index of gasoline dropped from 100 to 89, and that of fuel-oil from 100 to 86, while the general price index for the countries of the Organisation for Economic Co-operation and Development, OECD, rose from 100 to 140.

The rest of the present section will set out to examine three interrelated aspects or issues of pricing against the background just outlined. These are: (a) the economic influences that operate on the process of price determination; (b) the landmarks of price changes in the 1970s; and (c) the concept of a 'fair price' and what such a price ought to incorporate, with special reference to price adjustments since 1973. The mechanics and details of price changes, which usually occupy the centre of concern in discussions of oil pricing owing to their impact on the cost of oil to the various categories of consumers, will not detain us long. This is firstly because the story of changes in prices during the decade under examination, with the accompanying changes in relations between oil-producing countries and oil companies, is adequately told in the literature; and secondly because we intend to concentrate on the issues involved in pricing. Some of these issues have already been touched upon: we refer to the inter-connections between pricing, production, reserves, depletion, and conservation; some others will be discussed subsequently: the inter-connection between pricing, oil revenue, development, foreign aid, and the international economic order. We are left therefore with the three areas of inquiry just listed to explore here.

The *first* area, namely the influences that operate on the

process of price determination, must be explored because it is usually shrouded in confusion. It is important to ask whether this is deliberate or not, and some probing work has attempted to answer this question and to determine the motivation behind much of the deliberate confusion detected. Yet we will not take part in the polemics that such a probe involves, since we consider it more enlightening to dwell on those aspects of price-determination which are not the subject of much controversy and which can be supported by historical and analytical evidence.

Like all prices, that of oil is determined — in the long term as in the short term — by the interaction of supply and demand (that is, at the point of intersection of the supply and demand curves). But behind supply and demand lie many factors and influences which form their trends or the positions and shapes of the respective curves that represent them. It is these determining factors and influences that interest us in the present context. One very simple yet very pertinent and important observation at this stage in the discussion is that the analytical model relevant to the individual producing country (the counterpart of the 'establishment' or 'firm' within the 'industry' in textbook economics) is not a model of perfect competition but of oligopoly, no matter how small a producer the country happens to be. No producing country within the industry is so small by itself that its actions (volume produced, price charged) fail to influence the volume of the industry's supply and its price tangibly. Furthermore, neither the supply nor the demand for that one country's oil is vastly expandable at any given price in the short run, as the model of perfect competition stipulates for the establishment or the firm. Even for the industry as a whole, the expansion in supply (given a rising price) is possible only within the constraints of capacity, and these can be widened only after a long period of time measured in years.

There are other notable differences between the oil oligopoly and the textbook model of perfect competition. One of these is product differentiation. Oil is not a standard commodity, as the model stipulates, in view of the differentiation between crudes with respect to specific gravity, sulphur content, and other qualities, apart from location. (Hartshorn speaks of '130-odd crudes that OPEC countries export'.[42]) To go on with the comparison: the price of crude is not equal to marginal cost and marginal revenue at the point of the intersection of their curves. The price of oil throughout its recorded history has always been above this point of intersection, with a downward-sloping demand curve and an upward-sloping supply curve, irrespective of the position of these curves and the steepness or flatness of their slope, both in the short term and the long term. It is essential to remember in the present context that the situation was similar under the concessionary company regime (when the seven majors controlled and ran the oil industry) even more markedly than under the producing country regime (with 13 producers within OPEC, let alone other producers outside it).

Even during the decades preceding 1973 when the price of OPEC oil used to be determined by the foreign oil companies, there used to be a very large gap between the price charged per barrel and the cost of production — not just the marginal but also the average cost. This is particularly telling since the companies were their own clients and undercharged themselves for the oil which they bought for their own downstream operations within the closely-integrated system, both horizontally and vertically.

Today, this discrepancy is much larger. This is explainable by three influences. The first is that in the past, as it is today, price determination has to take into consideration many other factors as well as cost of production: indeed, this cost constituted then as it does now a very small component of price. (Even in the 1960s, the average cost of production in the countries of the Arabian Gulf used to be under one-tenth of the price charged.) The second is that the structure of the industry — or the 'model' in force — is such that it not only permits but can force into existence a considerable differential between the price charged and the cost of production. This was true when the structure consisted of a small group of companies acting in close co-ordination, as it is true today with governments acting in co-ordination. The third influence, though implicit in the first, deserves singling out. It is the vital and strategic nature of the commodity under discussion, the fact that it is non-renewable and is being depleted at a fast rate, and the versatility and unique qualities of this commodity which give it an advantage over any other source of energy.

A close look at price in its relation to the supply and demand for oil shows that the causal relationship is weaker than might be expected at first glance.[43] There is no special mystery that surrounds crude oil and makes the basic principles of its pricing different from those that operate in the pricing of, say, tractors or coffee. Yet there are peculiarities in the case of oil that are not encountered in most other instances. We have already referred to the fact that demand for crude oil is derived demand determined by the demand for refined petroleum products, and to the fact that oil is a depleting, non-renewable resource. These two facts influence demand and supply, respectively, in a special way. There are several other factors that give to oil a particularity that is not shared by many other commodities, whether natural or manufactured. The truth of this statement will be better appreciated if we recall the special characteristics: the versatility, and the technical qualities of oil (and gas) compared with the other sources of energy — qualities that make it easier to handle and to use in many sectors especially transport, and to use as feedstock in fertiliser production and the petrochemical industry.

One of the special factors on the supply side is the long lead time required to develop alternative sources of energy in sufficient quantity to replace oil on a large scale. Another is the environmental handicaps and hazards that have to be overcome if coal-mining and nuclear energy were to be developed in substantial measure; a third is the costliness of the development of non-conventional sources of energy. A further factor to mention is institutional: the subjection of pricing to the judgment of powerful sovereign bodies. These bodies, whether companies (as in pre-1973 days) or governments, take market considerations into account in determining prices; but they also take several other matters into consideration, including political and social factors of immediate and future relevance, in addition to purely economic factors.

This last point is of particular pertinence to pricing policies — Arab and non-Arab alike. The companies were no less concerned with the political interests and desiderata of their governments than the Arab governments are of their own countries. If the Arab governments pay much more attention to their long-term interests and therefore attach greater importance to the question of resource depletion than the companies did, it is because the governments have their present constituencies and the coming generations of their people to think about and provide for, while the companies were concerned with the dividends accruing to their shareholders (in the short and medium terms) whose interests were far from being identical or even parallel with those of the oil-countries' societies and economies.

Finally, in determining the price level of oil, the governments cannot but ask themselves a twin question: who benefits from an adequately corrected (that is, higher) price, and who from an unjustifiably low price? The obvious answer to the first part of the question is that the producing countries benefit from a higher price which gives them larger revenues to invest in badly-needed development and welfare. But more probing will also show that there is a global benefit to accrue to the larger pool of energy resources as a whole. This is so because a higher price for oil — given oil's pre-eminence among energy resources — would justify and promote the search for alternative sources of energy, the development of sources already discovered, and the resort to erstwhile marginal or even sub-marginal sources capable of development.

The answer to the second part of the question regarding the identity of the beneficiaries of an unjustifiably low price, is also obvious. It is the consumers. But, once again, more probing will reveal that such a low price would disguise the need for oil conservation on the one hand, and, on the other, would disguise the urgency of the search for alternative sources or the further development of sources already in existence. The overall conclusion which readily suggests itself, therefore, is that an adequately corrected price for oil, in line with the adjustments that pricing underwent after October 1973, is beneficial to producers and consumers alike, if the analyst's judgment is not to be short-sighted and narrowly-focused.

So much for pricing and the supply side of crude oil. On the demand side, too, one must look for special factors determining demand besides price. (This comes out clearly in the analysis of Janabi already referred to.[44]) One of these factors has already been mentioned: that crude oil prices are derived from the prices of final products. But the demand for these products itself is sensitive to prices in different ways, depending on the type of product and the type of user (whether it is the transportation sector, manufacturing, households, or the petrochemical industry consuming oil and gas as feedstock).

The prices charged to these various users diverge widely from the price of crude used in the production of the final products demanded by the users. Thus, in 1970, the price of crude oil FOB constituted only 8.02 per cent of the final price of an average barrel of refined products to Western European consumers; against this, freight, refining, distribution, dealer's share, and taxes accounted for 7.02, 5.60, 12.78, 8.57, and 57.91 per cent of final price respectively. Even by the end of the 1970s after a large measure of crude price correction had taken place, there was still a considerable difference between the average price of crude ($31.94) and that of an average barrel of final products ($54.09). The crude price FOB in 1979 accounted for 31.94 per cent of the price of the barrel of refined products, while the other cost elements listed above accounted for 2.14, 3.24, 18.82, 3.54, and 40.32 per cent respectively.[45] It is worth noting that taxes, both in 1970 and 1979, were the largest single component in the price of final products paid by the Western European consumer, that is larger per barrel than the price of crude.

As Janabi notes, it is the prices of the final products that influence the final consumer, not the price of crude used in the manufacture of these products. If so, therefore, then the oil companies buying crude from the producing countries will be influenced less by crude prices than appears on the surface, since the companies usually manage to absorb crude prices *and* make a comfortable profit over and above the price of crude and the subsequent costs involved in its transportation, refining and distribution. Once again, it would be instructive for the Western consumer to realise that he pays *more in taxes* per barrel to his government than he does for the crude to the producing governments, and that the Western oil company that supplies him with the refined products makes substantial profits.[46]

The partial insulation of demand against the level of crude prices to which the last paragraphs alluded is fortified by certain other factors which make the connection between demand for and prices of crude, or the price elasticity of demand, distant and weak. These include measures restricting the entry of crude into some countries, measures restricting usage, and indirect measures such as subsidies of various forms to alternative sources of energy or to research into such alternatives, conservation measures and regulations in major consuming countries, and the like. The policies, regulations and measures adopted 'are increasingly tending to be decided on with little regard to crude-oil prices. Therefore it seems logical to assume that long-term demand for OPEC oil is being planned independently of assumptions on its future pricing.'[47] If this analysis is accepted, and there is no reason to find fault with its historical underpinnings or logic, then analyses purporting to establish a strong tie between demand for crude oil and its price on the one hand, and price and cost of production on the other, lose a great deal of their credibility.

The *second* area to explore in the present section is the main landmarks of price changes in the 1970s. We will not go into detail in this narrative for the reasons given earlier, but will underline the vast changes in actual prices. The pressures that made for a drastic change in the level of prices and in the process of price-determination have just been discussed, while the third (and last) area of inquiry will involve the discussion of the concept of a 'fair price' of oil. For the present it is sufficient merely to stress the fundamental cause of dissatisfaction with the pricing realities of pre-1973 days. This was the very low level of prices (compared with comparable sources of energy), and their inability to allow adequate oil revenues for pressing national developmental and other needs in the producing countries, or to match the price rises of the large imports of these countries, particularly manufactured imports.

The OPEC countries had asked the companies to have prices linked to an index of prices of manufactured goods as far back as 1962, but this linkage was never effected. The malaise over crude price levels as set by the oil companies and their insensitivity to inflationary pressures characterised the 1960s in particular, since they witnessed an actual decline in the absolute level of prices, but the malaise grew noticeably as the decade neared its end and the 1970s arrived. Other grievances plagued the relations between oil governments and companies, though these in one sense or another were related to prices and/or government revenues. Indeed, as early as July 1963, when OPEC was barely three years of age, this organisation considered retaliation against the 'obduracy of companies', in the form of unilateral legislation regarding royalty expensing (that is, considering the royalties paid by the companies as a production expense instead of treating it as an offset against the governments' share in profits after expenses had been deducted from gross sales revenue). If this proved unsuccessful, the governments were to move on to such measures as imposing controls on oil production, exports, and prices, or even to partial nationalisation.[48]

As stated in the previous chapter, the 1960s were the decade of uneasy stirrings by OPEC's member governments, where the centre of disagreement or controversy with the companies moved from one issue to another. In this respect the 1970s were a clearer decade, for, apart from the events of 1970-2 in the area of pricing, the companies were moved out of this area by the bold decision of October 1973, stipulating for OPEC governments themselves to determine prices. From that date onwards, discontent with prices, by the governments or the companies, became like the discontent with the price of other major commodities in world trade, leading the one party to restrict supply or the other party to restrict demand, until finally a price level which both parties can live with (even if uneasily) has been established.

The decade of the 1970s started with one barrel of Saudi Arabian Light oil 34° (which is known as the marker crude, that is, the crude which forms the basis for the pricing of OPEC oil) being priced at $1.80 FOB. (It had been at $1.75 in 1951, rose to $1.93 in February 1953, rose again to $2.08 in June 1957, then dropped to $1.90 in February 1959, and to $1.80 in August 1960 — to stay unchanged at that level until February 1971 when it rose to $2.18.) But the decade ended with the price at $24 per barrel (from which it moved upwards by the first of January 1980 to $26).[49]

The rise in the price per barrel from $1.80 at the beginning of 1970 to $2.18 on 15 February 1971 was a very small one in absolute terms but of larger significance in relative terms. It came about as the result of an agreement reached between the governments and the companies in Tehran, at which all the accumulating and unresolved disagreements of the 1960s were settled, thanks to the determined stand of OPEC. As Seymour rightly noted, 'OPEC clearly moved with great caution in the 1960s, being prepared for interminable bargaining about what now look to be insignificant sums and showing great reluctance to use sovereign powers of legislation.'[50] Of significance also in the Tehran agreement was the stipulation of a floor price (tax-paid cost) that could not thereafter be changed by the companies unilaterally.

None the less, the price remained very low in absolute terms, considering the level of prices of imports from the industrial countries and the general, growing pressure of inflation,[51] but more importantly considering the wide gap

## New Policy Options in an Integrated Context

between the price levels of oil on the one hand and the much more expensive alternative sources of energy on the other, in spite of the numerous advantages of oil. It would be difficult to over-emphasise the lack of economic (or moral) justification for the stagnation of oil prices in the decades preceding October 1973, though it is easy to explain this gap by the wide difference in relative political and military power of the developing oil-exporting countries and the industrial oil-importing countries, which owned the oil companies.

Before we move on to discuss the significance of price changes in the very early 1970s, it will be necessary to outline very briefly the highlights of the concerns and actions of the Arab oil-producing governments (as OPEC members) with regard to pricing *and price-related matters* during the 1960s and the early 1970s down to the October 'turning of the tide':[52]

1. In their first conference (10-14 September 1960), OPEC members (then Iraq, Iran, Kuwait, Saudi Arabia, and Venezuela) 'expressed their serious concern at the continual erosion of their oil revenues because of actions by the international companies, and at the continuing decline of their oil resources. They resolved that any future price adjustment ... by the companies had to meet with the approval of oil-producing governments, that to ensure price stability the principles of market supply and demand had to be observed, and that the regulation of production had to be studied. The first resolution of the Conference established, as a principle, the legitimacy of the sovereign rights of oil producing countries to control the price of their crude oil, their only source of national income.'

2. In the fourth conference (5-8 April and 4-8 June 1962), OPEC 'protested against the oil companies' discriminatory action in unilaterally reducing oil prices. No effort had been made by oil companies to restore oil prices to pre-August 1960 levels, thus impairing OPEC Member Countries' purchasing power in relation to the increasing

prices of manufactured goods, fundamental and vital to their development ... The Conference also considered the formulation of a rational price structure to guide their long-term pricing policy with an emphasis on the linking of crude oil prices to an index of prices of goods which the Member Countries needed to import, to take account of the widening gap between the declining purchasing power of OPEC Government revenues and the increasing prices of imported manufactured goods and services.'

3. In the ninth conference (7-13 July 1964) 'OPEC Members recognized the need for a steady flow of oil to the international market on the basis of equitable and stable prices. This highlighted an immediate need to adopt, as a transitory measure, a production plan calling for rational increases in oil production from OPEC Countries to meet the estimated increase in world demand.'

4. The eleventh conference (25-8 April 1966) reflected its awareness of the resort by the oil companies to the manipulation of the level of production as a pressure mechanism on the producing countries, and stated its belief that such a manipulation was contrary to the national interests of OPEC members.

5. In line with their concern for the level of production, OPEC members 'reviewed the progress of the joint production programme' in the twelfth conference (4-8 December 1966). This concern derived essentially from the dissatisfaction with the low level of prices and therefore of oil revenues.

6. Again returning to the consistently bad price situation, the thirteenth conference (15-17 September 1967), in commenting on the position of Iraq and Libya with regard to the revaluation of posted prices, 'felt that action taken by the two Member Countries could be a prelude to a new era which could cause a break through the barrier of constant oil prices insensitive to favourable economic changes' − a reference to the 'prevailing healthy economic conditions in the international oil industry'.

7. The sixteenth conference (24-5 June 1968) adopted

an important document entitled 'Declaratory Statement of Petroleum policy in Member Countries', to which we had occasion to refer in the preceding chapter. The Statement emphasised the depletability of reserves; the connection between oil and development, and oil and government revenues and foreign exchange; the inalienable right of all countries to exercise permanent sovereignty over their oil resources; the importance of foreign capital and skills for development, but the resort to these as a supplement to national resources and manpower within the context of self-reliance; the necessity for government participation in the equity of companies; acreage relinquishment; and the determination of posted (reference) prices by governments. In one way or another, all the important themes of the Statement were related, closely or distantly, to prices.

8. The following conference (9-10 November 1968), emphasised the conservation of oil resources by insisting that the companies exploit the resources 'in conformity with efficient and rational methods to prevent waste and maximize yield.'

9. Once again, the nineteenth conference (14-16 December 1969) turned to conservation and gave its full support to Libya which had introduced production cuts for conservation purposes.

10. The twentieth conference (24-6 June 1970) 'adopted a production plan calling for a rational increase in OPEC Countries' production to meet the estimated growth in world oil demand during 1971-1975.' The statement went on to say that 'Recognizing that OPEC Member Countries were not drawing all the benefits they were entitled to from the exploitation of their hydrocarbon resources, the Conference stressed the necessity for the full integration of the petroleum industry in the national economies of Member Countries, through a systematic linkage between the hydrocarbon sector and the other sectors of economy.'

11. The twenty-first conference (9-12 December 1970) asked for an increase in the minimum tax rate on net income from 50 per cent to 55 per cent, and for a number

of other measures relating to price differentials for different types of oil and for different locations; it went on to declare that if the demands were not met, a procedure to enforce and achieve the objectives was to be determined. The conference once again noted and deplored the deterioration in the purchasing power of the revenue received per barrel compared with goods imported by member countries. It also demanded that prices be 'adjusted to reflect changes in the currency exchange rates of the major industrialized countries.' Finally, the conference 'supported actions taken by OPEC Governments over the unjustified slowdown of exploration and development efforts by the oil companies.'

12. In order to enable the member countries to implement the resolutions of the twenty-first conference referred to above, the twenty-second conference (3-4 February 1971) decided 'that each Member Country exporting oil from Gulf terminals was to introduce on February 15 the necessary legal and/or legislative measures to implement the objectives of the XXI Conference. Any oil company failing to comply within seven days would have all its oil and products shipments embargoed. The same measures were to be applied if the oil companies concerned did not meet the legitimate demands put forward by Algeria, Libya and Venezuela.' The oil companies complied, and the Tehran Agreement was signed on 14 February (one day before the deadline set) between the six Gulf Members of OPEC and 23 international oil companies. This Agreement was to last for five years. It settled outstanding issues with respect to price increases to account for inflation, increase of the tax rate by 5 per cent, the abolishing of oil discounts given by the companies and charged to selling price, and full royalty expensing.

13. The twenty-third conference (10 July 1971) went back to the question of a joint production programme, but only to hold its implementation 'in abeyance'.

14. The twenty-fifth conference (22 September 1971) asked the member countries to negotiate with the

companies in order to achieve effective participation, on the basis proposed by a Ministerial Committee which had been formed by the twenty-fourth conference (12-13 July 1971). 'In cases of negotiation failure, the Conference would determine the means of enforcing and achieving the objectives of effective participation through concerted action'. The conference also 'decided that Member Countries were to take the necessary action to offset any adverse effects on the per barrel income' as a result of the 'effective devaluation' of the US dollar. This subject was further studied at the twenty-sixth conference (7 December 1971), and a meeting with the oil companies was set for a discussion of the dollar issue. The conference also examined the question of the implementation of the principle of participation.

15. The meeting to consider the devaluation of the dollar was held on 20 January 1972, in Geneva, and agreement was reached for an increase in prices to compensate for the devaluation. The 'US dollar's movement was to be pegged to exchange rate fluctuations of a "basket" of major international currencies ... This agreement made provisions for further adjustments in oil revenues between 1972 and 1975.'

16. The thirtieth (Extraordinary) conference (26-7 October 1972) marked the implementation by some OPEC members of an agreement reached earlier on participation.

17. The thirty-second (Extraordinary) conference (16-17 March 1973) 'examined the world energy situation ... It agreed that access to the technology and the markets of the developed countries for the future industries of Member Countries, together with a just valorization for their hydrocarbon resources and adequate protection of their revenues, were essential objectives of the Organization. It was also agreed that the exploitation of petroleum and its trade was to be linked to a process of rational and accelerated economic growth ... The change in the value of the US dollar and its adverse effect on the purchasing power of oil revenues of Member Countries was also reviewed, and a

Ministerial Committee was set up to discuss the matter in detail.' This Committee subsequently held meetings with the oil companies, and finally, on 1 June 1973 the Geneva II Agreement was signed, providing for further compensatory price increases to make up for dollar losses.

18. The thirty-fourth conference (27-8 June 1975) issued another important policy statement in which OPEC asserted that it 'should not only try to attain the appropriate value for its oil, but should establish and foster permanent and diversified sources of income ... It was indisputable that the oil flow from OPEC Member Countries had contributed substantially to the development of the industrialized countries. As a reciprocal gesture by the industrialized countries, and to ensure a regular and steady flow of oil, they were expected to respond to the inadequate economic conditions to which most developing countries were subjected.' The Statement went on to indicate the course that co-operation between the two groups of countries should take, and ended by indicating the intention of OPEC member countries to 'further strengthen cooperation with the oil importing developing countries whose energy requirements were increasing.'

19. The final conference before the meeting of the Gulf members of OPEC in October which decided on the takeover of pricing by the sovereign states themselves, was the thirty-fifth (Extraordinary) conference (15-16 September 1973), which decided that 'the level of posted prices and the annual escalations provided for [by the relevant agreements in existence] were no longer compatible with prevailing market conditions and galloping world inflation.'

The terms of these agreements were to be revised. But events overtook negotiations for revision, when on 16 October, the structure of relative power between the oil governments and companies was dramatically, radically, and irreversibly altered. From October 1973 onwards, discussion and agreement on the marker crude price was no longer a matter

*New Policy Options in an Integrated Context*

between OPEC members and oil companies, but among the members themselves. However, the six or seven years of the decade after October 1973 were not a continuous honeymoon of mutual understanding and harmony among OPEC members, or among the Arab members by themselves. Indeed, a great deal of misunderstanding and friction, arising from differing positions of principle and viewpoint, erupted, leading for a while to a 'two-tier structure of prices' – that is, two levels of prices applying simultaneously in the case of the two groups of countries in dispute about the appropriate level to impose. It must be added that such differences with regard to pricing have also more than once been associated with differences with regard to the optimum level of production to be sought. The following continued chronology of events and decisions pertaining to prices or price-related matters will reflect the differences, as will the discussion which will follow the chronology.

20. The price of the marker crude was raised by unilateral government action by 70 per cent on 16 October (from $3.011 to $5.119 per barrel). It was to be raised again to $11.651 at a meeting of the Gulf oil ministers (22-3 December 1973), effective 1 January 1974. This price, it was believed, would bring the level nearer the point where it would compensate for inflation and dollar decline versus the major currencies.

21. The new price was to remain in effect until 1 January 1977, at which time new increases would become operative. The price freeze from 1974 to 1976 was insisted upon by Saudi Arabia, supported by the United Arab Emirates, although the price level was still well below the correction which inflation and decades-long under-pricing justified. Indeed, the official price of the marker dropped between 1 January 1974 and 31 December 1976. The slight increase as of 1 January 1977, from the official price of $11.51 to $12.09 per barrel (the former operative from 1 October 1975) was set in the forty-eighth conference (15-17 December 1976) during which the split over prices

already mentioned occurred. The conference, except for the two countries mentioned, decided to apply a 5 per cent increase till the end of June, and another equal increase as of July 1977. The two countries excepted decided to apply one increase only.

22. However, it was announced on 9 June 1977, that the Arab members of OPEC except Libya agreed to forego the second 5 per cent increase for the sake of unity and solidarity within the Organisation, but Libya was to follow suit later at the forty-ninth conference (12-13 July 1977). Thus price reunification was achieved.

23. The oil ministers set up a Committee on Long-Term Strategy (at an informal meeting, 6-7 May 1978), 'to examine future strategies for the Organization and the policies which it should follow in the medium and long-term.'

24. Once again in mid-1979 OPEC failed to reach a unified policy with regard to prices. This was caused by the feeling of the majority (with the notable exception of Saudi Arabia) that the inflationary pressure coupled with the disruption in and decline of oil production in Iran (subsequent to the revolution and the overthrow of the monarchy) justified a tangible price increase. Thus, the fifty-fourth conference (26-8 June 1979) decided 'to adjust the Marker Crude price to $18 per barrel; to allow Member Countries to add to the prices of their crude a maximum market premium of $2 per barrel over and above their normal differential, if and when such a market premium was necessitated by market conditions; and that the maximum prices that could be charged by Member Countries were not to exceed $23.50.' The setting of a price ceiling was thought to provide partial satisfaction to the discontent among the majority over the price set and to allow for flexibility in pricing to individual governments, provided their prices did not shoot through the ceiling. (Saudi Arabia kept its price at the $18 level till November.)

25. The fifty-sixth (Extraordinary) conference (7-8 May 1980), concerning oil supplies, 'reaffirmed the intention of

Member Countries not to make up losses of Iran's oil exports in the international oil market.' (However, Saudi Arabia's production averaged 9.9 million barrels per day for 1980, as against 9.5 million for 1979 and 8.3 million for 1978. The official explanation was that the increase in production was motivated by the desire to force the price down, which in turn was motivated by the country's concern for world economic growth.)

26. Again, the fifty-seventh conference (9-11 June 1980) in order 'to stabilize the international oil market, decided: to set the price level for a Marker Crude at a ceiling of $32.00 per barrel; that the value differentials which would be added over and above this ceiling for the Marker Crude, to take account of quality and geographical location, should not exceed $5.00 per barrel; and that this price structure was to be applicable as of July 1, 1980.'

27. The last decision on prices to be taken before the end of the second decade of the life of OPEC, like the one just reported on, extends beyond the period covered in this book. But it is appropriate to include both decisions taken in 1980 because 1980, like 1979, was a year which witnessed strong upward market pressures on prices. (The year 1981, in contrast, was a year which witnessed a marked slackening of demand and downward pressures on prices.) The fifty-eighth (Extraordinary) conference (17 September 1980) 'decided to fix the price of the Marker Crude at the level of $30 per barrel and to freeze the other official prices of OPEC Member Countries' crudes at that level until the next meeting of the Conference.' (Again, Saudi Arabia chose to keep the price at $28, and not to raise it to the level set by OPEC. The divergence between the Saudi price and that of other members for the marker crude was to continue (though the price level of both groups rose again) till finally price reunification was achieved at the sixty-first (Extraordinary) conference (29 October 1981), when the price of the marker crude was set at $34 per barrel where it was to stay unchanged until the end of 1982. By the time of reunification, there was a

wide gap between the then prevailing Saudi price of $32 and that of the other members of $36 per barrel of marker crude. The $34 level was a compromise finally agreed upon unanimously. In addition, Saudi Arabia, in a 'separate yet obviously related move, announced an output cutback to a ceiling of 8.5 million b/d... with effect from 1 November' — or a cutback of about 1 million b/d and a 'restoration of the traditional 8.5 million b/d Aramco output ceiling which was enforced before the onset of the 1979-81 oil crisis.'[53]

The chronology just traced and the controversies within OPEC after October 1973, mostly with regard to crude price levels but also to volume of production, justify some explanation. This would help clarify pricing policy, both Arab and non-Arab, inside OPEC. (In the present context, the main parties to the formulation of such policy were the Arab countries, because the preponderance of their aggregate production wields a deservedly large proportion of power inside the Organisation. But this is not to deny the very prominent influence of Iran and Venezuela and the significance of the positions they usually took in the shaping of pricing policy.) Several observations will be made with a view to describing and analysing the course of prices, and explaining the background to pricing controversies.

1. The first observation is factual. It is the sharp contrast between oil price level in the 1960s and the 1970s, and between the courses of prices for each of the decades. Thus, while posted prices in the 1960s were generally below $2 per barrel FOB for the marker crude, and they remained unchanged throughout (with realised prices some times even dipping below posted prices), they witnessed a steep increase during the 1970s, reaching $24 per barrel at the very end of the decade, with the years 1974 through 1979 registering the most notable gains.

2. The 1970s by themselves can be divided into two phases. The first can be designated as the phase of mild price correction 'within the system' — that is, within the company regime;

this phase stretched from the beginning of January 1970 to the middle of October 1973. Consequently, the corrections undertaken were very marginal in absolute terms, though they looked a little more important in relative terms. More specifically, they were more real with respect to compensation for dollar devaluation, than they were as compensation for past underpricing plus inflation. Indeed, they barely exceeded the inflation experienced between the beginning of 1960 and 16 October 1973, the day of the take-over of control by governments, but left unaccounted for the gross underpricing of the past decades.

The second phase included price corrections within the era of control, that is, since mid-October 1973. The first such correction involved a 70 per cent increase on 16 October, but this was soon recognised by OPEC members as glaringly inadequate, though justifiable as a first step. The second correction became operative as of 1 January 1974; it brought the price of the marker crude to about 4 times its level on 15 October before the first correction, but 2.3 times the price set on 16 October. The new price of January 1974 was generally thought of as representing a fair approximation of the level that the marker price should reach if it was to nearly allow a reasonable correction for past underpricing and to comprise a reasonable compensation for old inflation and newer and stronger inflationary pressures in the early 1970s. However, price was to be greatly outdistanced by 1979, and as a result the price that was operative by the end of this year was $24 per barrel, or twice the level at the beginning of 1974. This seeming escalation in the crucial six years 1974 through 1979 needs careful examination.

3. The first three years of this period witnessed a decline in official government selling prices in current money, but more so in real terms. Oil prices were able to pick up their ascent tangibly only in the middle of 1979, after three or four years during which the oil-exporting countries suffered a tangible loss — estimated variously at between 20 and 30 per cent — of the purchasing power of their oil dollars. This erosion can be established whether one uses the index numbers of OECD

export prices or those of OPEC import prices, allowing for minor adjustments called for by the fact that export prices are recorded FOB while import prices are CIF.[54] In fact, it was only with the price increase that became operative at the beginning of 1980 that OPEC succeeded in stopping the erosion of real prices.

Behind the unevenness in the course of prices from 1974 through 1979 lay a basic difference in position within OPEC with regard to pricing. Most of the members, with the notable exception of Saudi Arabia (with whose position the United Arab Emirates associated itself) believed the price set as of January 1974 to be fair and justified by all considerations. But Saudi Arabia believed that the adjustment of 1974 had gone beyond the absorptive capacity of the market. Unable to bring about a price reduction satisfactory to itself in 1974 (apart from a minor one that was to come in November), Saudi Arabia resorted to expansion in production in order to put tangible downward pressure on prices. Thus, it produced in 1974 almost 12 per cent more oil than in 1973. Prices were placed under pressure and by January 1975 the level was lower by 7 per cent than it had been two months before. Demand was as brisk in 1974 as it had been in 1973, but it declined considerably in 1975, and the decline was reflected in production (which was almost 12 per cent below that of the year before). The decline, however, was short-lived, since 1976 and 1977 registered first a return to the level of demand of 1974, and then an excess over it.

Looking back at the years 1974-7, many observers and analysts are convinced today that the failure of real prices to keep their purchasing power during the years in question made necessary the big jump of 1979, even though other factors were active (predominantly the revolution in Iran). The moral in this is that, as in the case of price adjustments in the autumn of 1973, 'unsettled bills' with regard to gradual and justified price increases over the years bring about sudden and sharp increases later.

4. The utilisation of production by the major OPEC producer, namely, Saudi Arabia, as an instrument to influence

prices has unique features. On the occasions that production levels have thus been used, namely in the period under examination and subsequently in 1979-81, this use has been diametrically opposed to what it would have been had OPEC acted as a typical cartel. Thus, the leader of the so-called cartel (which the opponents of OPEC insist on calling it) uses the volume of its production as a pressure mechanism *not* to raise but to depress prices.[55] If OPEC is a cartel, then it surely is a very atypical one, whose leader is interested more in accommodating the consumers' demand for lower prices, than that of its fellow-producers for higher prices.

The willingness of the leader to use prices and/or supply as a pressure mechanism in the face of its recalcitrant associates has been at the heart of the controversy now and then erupting within OPEC with regard to prices charged and supplies produced. Saudi Arabia has resorted to supply movements in order to maintain a price level considered satisfactory or to bring down a price level considered unsatisfactory. But it also used its own lower price quotations for pressure, when the two-tier price system prevailed and again during the 1979-81 price controversy. Producing by far the largest part of OPEC's production, its pressure on the market cannot be overlooked. (Since 1973, Saudi Arabia's production has ranged between one-fourth and one-third of total OPEC production, but has averaged about 45 per cent of Arab production.)

Two main arguments have usually supported Saudi pricing policy which has generally been more moderate than that of most OPEC members, particularly during the period 1974-7, and later during 1979-80. The first is that the market does not tolerate the price increases decided upon by the majority, and the second that world economic growth (and the economies of the developing oil-importing countries) required gentle price increases when upward adjustments were at all justified. The logic behind this position was that world economic prosperity would be in the interest of the oil exporters as well. The second argument should be well taken, although concern for the welfare of other regions in the world, particularly that of the industrial countries, should be mutual and not one-sided. But the strong inflationary pressure 'exported' by this powerful group of countries is no evidence of such concern by OECD countries, either where Saudi Arabia or where other oil exporters are concerned.

However, it is the first argument that raises serious questioning. This is because the claim that market forces of supply and demand for oil press for a drop (or a non-increase) in prices, cannot be convincing when these same forces are influenced by policy decisions purporting to press downward on prices. To expand production considerably when the demand is slack, and to keep the price of the leading producer low at the same time, cannot be construed as submission to market forces pressing *spontaneously* on prices. This is particularly so when price and production measures are taken alongside or even ahead of the presumed 'spontaneous' market pressures.

It is clear that there is here a wide divergence between the logic and processes of pricing, as perceived by the industry's leader, and by the other Arab oil exporters. This divergence also emerges with respect to the volume of production, which Saudi official sources have repeatedly described as far in excess of what Saudi national needs by themselves would suggest.[56] The explanation given for the adoption of such a volume is concern for the international economy – that of industrial and developing oil-importing countries alike. It is necessary to point out here that the proper explanation of the divergence of opinion between Saudi Arabia and the mainstream of Arab exporters is not to be found wholly in claims of excessive virtuousness and predominant concern with the interests of the world economy, as the supporters seem to attribute to Saudi oil policy with respect to production-cum-pricing; nor is it to be found in the accusations of subservience to Western interests and non-concern with Arab interests, as the adversaries claim.

As far as the mechanics of the use of pricing and production referred to are concerned, a pattern has been established whereby Saudi policy pursues price moderation plus expansion of production in a tight market, and price reduction plus *expansion* of production in a slack market in order to force the other producers to lower their prices as well, with the declared object of activating demand. The other producers prefer a policy aimed essentially at rising prices in a tight market plus expansion in supply, but stable prices and a *reduction* of supply in a slack market. In brief, the Saudi preference is for a more-freely fluctuating price level, while that of the majority of other OPEC producers is for a price level that could rise but should generally be shored up through production restraint or even decline when it is under downward pressure.

The explanation of the difference in policy and in the reasoning behind it can only be found in a delicate mix of factors, not in one over-simplified factor. In the present writer's view, and there is sufficient evidence to support it, the production-cum-pricing policy under examination is to a considerable extent, but not wholly, guided by Saudi Arabia's conception of its sense of international responsibility and by the strength of its ideological and political affinities. But there are four other policy determinants of great strength. These are, first, concern for the health of the United States dollar in which prices are quoted and oil revenues accrue; secondly, concern for the health of the United States economy where Saudi Arabia has invested or placed the vast bulk of its oil revenues not used at home; thirdly, concern for the country's international image as a 'moderate and reasonable state', and concern for the country's ability to acquire strong political leverage, particularly with the United States; and, finally, the fact that the country does not have to operate under financial or resource pressure – given its huge oil reserves and financial reserves – and therefore is not as exposed as the other exporters are either to the advocacy of steep price increases or to that of harsh conservation measures. Having said all this, it remains true, it seems to us, to say that while such reasoning is basically correct in general, it can be, and has been, carried too far. It is the over-loading of this reasoning which at times occasions the accusations referred to in the preceding paragraph.

5. The final observation to make with respect to the second area of inquiry in the present section of this chapter is that the pricing of crude has not been tied to a pricing index (however the index is constituted), in spite of the repeated references to indexing, and the pressures by most oil exporters for it. The most prestigious body to give credence to the idea of indexing has been the OPEC Ministerial Committee, set up in May, 1978, to suggest a long-term strategy for OPEC which would guide the Organisation in achieving its objectives in a world that has changed vastly since 1960, when OPEC was established. This Committee drew together a group of experts that produced a report which was submitted to and subsequently approved by the Ministerial Committee in its fourth meeting (21-2 February 1980).[57] Among other things, the long-term pricing strategy emphasised the need to relate prices to indexes which take account of inflation and fluctuations in exchange rates. Prices were also to be adjusted in a manner that permitted them to reflect real economic growth in the industrial countries. (The relations with these countries as well as the developing countries formed other areas covered by the recommendations.)

However, this Report has never been given the supreme sanction of OPEC, whose members have had widely-diverging views concerning its approval. There are already references to 'changing circumstances' that call for a reconsideration of the whole Report. The indications are clear, from reports in the professional journals, that the subject of indexing, like the rest of the Report, is destined for another extended period of dormancy under pressure from certain moderating influences in the area of crude oil pricing. The subject will be discussed no further here, since indexing has not formed a part of OPEC's pricing experience so far. But the rationale behind it will come up for some discussion in the context of the discussion of the 'fair price' of oil, to which we will turn immediately.

The *third* area of inquiry in this section of the present chapter attempts to identify the components of a 'fair price' of crude oil, or to identify the principles and factors which are capable of determining the level at which the price can be considered fair and reasonable by a responsible, objective observer. One is reminded, at this point, of Hartshorn's biting, but largely true, statement that the oil exporters still feel they have to justify their prices and seek 'consent from the other side of the table' as though they were trade unionists bargaining with management. He views the situation differently. To him these exporters:

> are now traders putting a price on something that customers want, take it or leave it (in the ground). This is not yet always manifest in some of their spokesmen's arguments. They often still seem to be seeking the customers' approval as well as their extra money — a vote of thanks as well as a higher price? But gradually that is being replaced by greater commercial self-confidence, as befits market power.[58]

Be that as it may, it seems to us justified to explain why it has been necessary for the oil exporters to make a substantial correction of prices since mid-October 1973 that goes well beyond the inflation within the period. Such an explanation can therefore be considered neither as an apology nor as a reflection of lack of Arab self-assurance with regard to oil pricing. At the minimum, an explanation would remove a great deal of confusion, and even of bitterness, which arises as a reaction among oil consumers whenever the question of oil prices is raised.

It was stated earlier in this section that the price of crude oil is a derived price, which is substantially determined by the demand for refined products and the prices which they fetch in the market. But this is not quite a one-way relationship, since as the price of crude rises, other things being equal, its 'responsibility' for the prices of refined products increases. Consequently the analyst would not be justified in saying that crude prices can be raised as high as the demand market for refined products can bear, and leave it at that, any more than he would be justified in saying that, since the price of crude is a residual (after refining, transportation and distribution costs and taxes have been deducted from the price of refined products), the producers will have to accept whatever price this residual amounts to. This would be tantamount to giving the sovereignty over crude prices to the refiners, the tankers' owners, and the consuming country's taxing authorities by depriving the producers of such sovereignty.

What we are leading to is that the producers must be assured a fair price which includes the components being identified here, and that the refined products must bear the costs necessary in their production and transportation and storage. Whether or not taxes are added by the consuming governments is a matter for them and their constituencies to settle. But the producers cannot be held responsible for the final price the final consumer has to pay, if his government chooses to add a tax component to the tax-free sub-total of costs, larger even than the price of crude per unit of oil. Nor would it be right to blame the producers for insisting on a fair price for their crude which accounts for the various elements that ought to be included.

The price of crude has to be capable of being analyzed and defended, partly for the consumers who need to understand the grounds on which a fair price is based, and partly for the producers themselves who must find out if the various components of the price (or the factors in pricing) have been reasonably accounted for, with a minimum of 'sins of omission and of commission'. In the present writer's view, it would be neither sufficient nor fair merely to say that the price set is dictated and permitted by the interplay of the relative power of the two sides to the bargain, the producers and the consumers. The former are under no legal or economic obligation to explain and justify the prices they charge, but they are under a moral obligation to explain their decisions and make them convincing. This analysis is motivated by this conviction.

Many price components have already been mentioned in

one relevant context or another; some have not been. The purpose of the present discussion is to consolidate and systematise the listing of components, and to include brief explanations as the need arises. It ought to be remembered all along that essentially the discussion attempts to explore the factors that lie behind the supply schedule of oil, but to a certain extent also those behind the demand schedule insofar as producers take into account buyers' preferences and the forces that create these preferences, as one element in determining the level of prices they ask for. Therefore, while it is obviously true that what determines price is the interplay of supply and demand, this does not say much, if anything, about the forces that determine supply and demand themselves. Consequently the probing for a *fair* price, which we shall do presently, is no more than the attempt to identify those forces which constitute and shape supply in the producers' mind, and the conception by the producers of the forces that constitute and shape demand.

1. The first and most obvious component of the price we are attempting to dissect is the cost of production of oil. It is the average cost which it is appropriate to account for, since it would include the share of every barrel in the fixed costs involved in the exploration and development of oilfields up to the point where the oil starts flowing, as well as the cost of current production. In contrast, the marginal cost would be misleading because it would only account for the share of the marginal barrel produced in the inputs used directly in its production. Thus, the cost of production would be the basic component in the price to be charged, to which several other components should be added.

2. While it is relatively easy to compute the first component, the next to include would be extremely difficult. This is a premium or allowance for depletion of present reserves, considering the fact that oil is a non-renewable resource the life of which the producers wish to extend. The rate of depletion permitted, given the proven reserves, is a function of the higher policy of the government concerned.

This policy, in turn, is a function of the level of development achieved in the economy; the stretch of the process of basic development envisaged; the oil revenues estimated to be required for this process; the availability, sources, and cost of alternative energy once the oil is depleted; and, in the international context, the extent to which the producing country feels inclined to accommodate the oil needs of the importers. The inclusion of this last element for consideration cannot be dissociated from the producers' evaluation of the degree to which present oil consumption levels in the importing countries are essential or wasteful, and their evaluation of the extent to which the importers can (and wish to) substitute other energy sources for oil. While it is not up to the producers to determine the level and style of life of the oil-importing country, it is within their right to pass judgment on whether they justify the sacrifices the producers have to undergo in providing the non-renewable oil resources to maintain them. Obviously, if the consumers do not apply self-discipline in their consumption of oil where such discipline is called for and possible, the producers are justified in expressing their disapproval through the price they charge for their oil resource. On the other hand, to the extent that their expectations of future finds of oil (or their ability to improve recovery rates) are bright, the producers can slacken the conditions for export with regard to the volume of production and/or the price they decide upon.

So far, the analysis has proceeded as though there is one producing country only. But reality is different. OPEC speaks for a community of producing countries, all of whom are sovereign states. Consequently, in taking the depletion component of pricing into account (even if concern with depletion is not externalised but merely borne in mind), the Organisation does so for the community as a group, as though the individual countries' reserves formed one whole pool, in spite of the widely differing endowments among the member countries. Consequently, the position OPEC takes on the need for prices to take account of depletion is one reached by consensus, although the depletion component in the price will in fact vary considerably from one case to another. Thus, Saudi decision-makers bear it in mind no less than those of a much less-endowed country like Algeria. (However, in fairness it ought to be added that the Saudi tendency to be less demanding on price increases within OPEC must be to some extent influenced by the much greater abundance of the country's reserves and its ability therefore to deplete these reserves over a much longer period of time than, say, Algeria. This point was referred to in the latter part of the discussion in the second area of inquiry in this section.)

3. Closely related to the depletion component is the incentive component. Thus, the price must include an element that can serve as an incentive for further exploration and development of new oilfields, one which is capable of 'justifying investment for the addition of new higher cost reserves'.[59] Furthermore, the incentive must also aim at encouraging enhanced recovery from the oilfields already in existence. This element is important since the oilfields already developed have great promise of future output if improved, but costly, methods of recovery are used. And it can readily be justified not only from the point of view of producers, but also from that of consumers in whose interest it is to expand the supply of oil resources, given the advantages which oil enjoys among energy sources.

4. The superior qualities of oil and its great versatility also justify a premium, compared with alternative sources of energy, even were the other grounds for comparison to be more or less equal. Thus, a unit of oil cannot be priced equally with a unit of oil equivalent of some other energy source, in disregard of the multiplicity of uses to which oil can be put, whether as fuel or as industrial input and feedstock. Indeed, a premium to allow for this technical superiority, convenience, and versatility would be particularly useful in order to restrain the consumption of oil as a fuel, and to conduce to the conservation of as large a proportion of its reserves as possible for 'nobler' uses for which no other energy resource is equally usable. Such

conservation must be welcome to consumers as much as to producers, since both groups wish to extend the life of oil as long as possible.

5. The pricing of crude cannot be determined in isolation from the prices of alternative sources of energy, even with the various premiums and allowances taken account of. In other words, there is a strong case for the price of oil going up until it approximates to the price of alternatives. But before we comment on this factor in pricing, we must indicate that the approximation is much easier talked about than achieved, owing to several qualifying points that have to be made. Some of these will be stated:

> Two of the alternative sources of energy, coal and nuclear power, are today cheaper than oil per barrel of oil equivalent. However, though cheaper, their expanded production faces stiff resistance on social and environmental grounds, and the theoretical advantages of these sources are beset by practical disadvantages. (World coal consumption has increased by 2.1 per cent annually on the average between 1970 and 1980, as against 23.7 per cent for nuclear power. However, consumption of nuclear power still amounted in 1980 to only 167.4 million tons of oil equivalent (against 3,001.4 million tons of oil), compared with 19.8 million tons and 2,281.7 million tons, respectively, for 1970.
>
> Neither coal nor nuclear power can be put to the same uses as oil; in other words, substitution between each of them and oil is very limited because of technical particularity.
>
> The other alternative sources of energy, whether primary or synthetic, are on the whole substantially costlier than oil to produce and place on the market. Furthermore, there is no uniform costing for the development of the various alternatives to guide policy-makers in their decisions, owing to the wide range of prices which the alternatives would demand once developed.
>
> To choose one alternative rather than another as a

standard for comparison requires, apart from information on cost and price, that the alternative chosen should potentially be available commercially, once developed, on a scale sufficient to substitute for oil if the need arose. An alternative which meets the cost and price requirements under laboratory conditions need not necessarily be right for comparison with oil owing to its limited eventual availability to the market.

The problem of limited substitutability is relevant here, as in the case of coal and nuclear power. Each of the alternative sources of energy can be a substitute for some oil products; none can be a complete substitute, especially where it comes to gas and certain refined oil products used for transport, or as feedstock in petrochemical industry.

Looking at the whole question of alternative sources of energy, the logic of the situation suggests that OPEC should concern itself with policies relating to the development of alternative sources, as part of its concern with future oil policies.[60] But as far as pricing specifically is concerned, it would seem to the present writer that raising oil prices until they closely approximate the prices of the nearest-priced, nearest effective substitute must only be done with caution. This is because the price increases within the ceiling set by the prices of potential alternatives, must not reach a point where demand is inhibited from clearing the export market of the supply that the producers need and want to sell, given their depletion/conservation, and their development/consumption preferences. Furthermore, the approximation of the prices of effective substitutes must be achieved gradually, within the absorptive capacity of the market, and with concern for economic growth in the various regions of the world. On the other hand, the oil-importing countries, particularly those within the OECD, should show at least equal concern and consideration for the interests of the oil-exporting countries, to whose restraint and sense of responsibility the economic well-being of OECD, and the energy situation as a whole, have owed and still owe a great deal.

6. Another consideration in the constitution of a fair price for crude oil is compensation for world inflation and currency fluctuations. It is admitted now by objective analysts that the increases in oil prices were in part meant to compensate the producers for inflation imported from the advanced industrial countries, and were not a primary cause of this inflation. This was evident when the first notable increase took place in mid-October 1973, when inflation had already been substantial between the beginning of 1970 and the end of September 1973.[61] It was also evident when after the initial adjustments in October and December 1973, the price of oil actually declined somewhat in current terms between 1974 and 1978, but declined by 20-30 per cent in terms of purchasing power.

The Arab oil-exporting countries are concerned that their dollar oil revenue should at least maintain its purchasing power, that the terms-of-trade between their one unit of export and their imports should remain in balance. It ought to be borne in mind in this connection that only a part of the increases of crude oil prices since October 1973 has been to compensate for inflation. Another (small) part has been applied to compensate for the fall in the value of the dollar; but a significant part has been meant to correct the underpricing of oil in the 1950s, the 1960s, and the first few years of the 1970s.

Fluctuations in Western currencies, particularly the dollar (in which prices are denominated and in which the largest single portion of the oil countries' reserves are kept) coupled with inflation, have cost the Arab countries heavily in terms of the erosion of the purchasing power of their reserves.[62] They see no reason why they should expand their production to the point where excess revenues are accumulated and built up abroad as reserves, only to see the real value of these reserves eroded by inflation and also by currency depreciation. What adds to the dissatisfaction of the Arab oil exporters is that the rise in the prices of their imports from OECD countries — whether capital goods, consumers' goods, or technological software and other services — is higher than the inflation rates concurrently experienced within the exporting countries. This, plus indications of price discrimination exercised against the oil countries compared with other importers, justifies in the view of the former the insistence that oil prices must run parallel with the course of OECD export prices as these are charged to importers in the Arab oil countries.

7. Fair pricing also requires that the price of oil should bring in revenues that are badly needed for development, and that this development should be aimed at within a reasonable time horizon during the lifetime of Arab oil resources. This means that prices should be high enough — within the constraint of the ceiling of the prices of effective substitutes for oil — to permit the accrual of the desired volume of revenues, given the length of the development process and the finiteness of oil resources. No serious analyst would pretend that the right formula for pricing, given these conditions and constraints, is easy to derive. But it is not difficult to appreciate the trend and content of the reasoning involved. The issue is of particular insistence in view of the niggardliness of nature in many of the Arab oil-exporting countries and their utter dependence therefore on oil revenues to be the lubricant of development, and on the oil sector (particularly the industrialisation of this sector) to be the engine of development.

8. Finally, a fair price for oil should of necessity move upwards (always within the ceiling set by alternative sources of energy) for one very important internal reason and another very important international reason. In the latter instance, a rising oil price would not only justify but also permit the active search for and development of alternative sources of energy, which the world at large should pursue in order not to have a strangling energy crisis in the decades to come. From the Arab point of view, a rising price which leads to discipline in oil consumption (domestically and abroad) and to a slowing down of depletion is essential for strategic and economic security reasons. The owners of oil resources, and the Arab region as a whole (which forms the strategic depth

of each of the oil countries) must not allow themselves to be exposed to excessive dependence on the industrial countries. This they would do by dissipating their hydrocarbon resources quickly and subsequently finding themselves largely at the mercy of these same countries that would have developed alternative sources, after having mostly burnt up Arab oil (and gas). Internally, the Arab region will be well advised to remember that its era of intensive industrialisation is still ahead of it, and it would be a blessing, when this era sets in, if the oil and gas which the region needs are still there in abundance underground.

The presentation and discussion of the eight components or elements of a fair price for oil does not constitute a formula for the setting of a price in dollars and cents. The fact that some of these components defy quantification does not mean that monetary values cannot be attached or imputed to them — some sort of shadow values as are resorted to in social accounting. But the weight of our argument is placed on the inescapable need to bear in mind and take account of the various components or considerations that must enter the constitution of a fair price for crude oil. That some of these are not always included, or if included are not always externalised, does not reduce their relevance and imperativeness. To keep sight of the principles and components of fair pricing would remove a great deal of the confusion that has attached itself to oil pricing in the 1970s, and of the misunderstandings and recriminations that have resulted from the confusion — whether spontaneous or deliberately-created by the opponents of OPEC. But to lose sight of the principles and components of fair pricing would in the long run prove neither in the interests of oil exporters nor of importers — the decision-makers who in the final analysis shape the supply and demand curves for oil which determine its price.

*New Policy Options in an Integrated Context*

## Downstream Operations: Policies and Implications

A discussion of policies relating to the hydrocarbon sector — including both oil and gas — would be incomplete if it was confined to upstream operations and stopped at the point where crude oil was produced, priced, and marketed, particularly if it were to place oil policies within an integrated context, as the title of this chapter suggests. Certain other operations — and policies — of great significance and relevance have to be included; these are grouped under the term 'downstream operations', and they consist of the transport of oil (and gas) by pipeline or tanker, gas processing or treatment, oil refining, and petrochemical and other oil and gas-related industries. Except for the transportation of oil and gas, the exclusion of which was justified in Chapter 1, these will be examined in the present section.

It is essential to consider the policies that govern these operations, essentially because an economy's utilisation of its hydrocarbon resources would be sub-optimal unless it comprised the development of the resources beyond the production and export of crude oil, as well as the control by the society of subsequent advanced processing of oil and gas and their transformation through industrial activity. This can be achieved in its fullest and most defensible form, if the downstream operations themselves were inter-related and integrated in a manner which permitted gas treatment and oil refining to feed the industries that use oil and gas directly and heavily, both as source of energy (fuel) and as feedstock (raw material). This integrated approach has not always characterised the policies governing downstream operations, but it has gained considerable acceptance, if not necessarily implementation, among the Arab oil-exporting countries. The compelling reasons for such an approach will become evident as the discussion proceeds; for the time being we will merely state that the three components of downstream operations will be examined together in the present section, although we will first deal with them one by one within the section, as their particularities and as clarity of discussion require.

Three observations are necessary before we proceed with the discussion. The first involves the admission that the decade of the 1970s with which we are primarily concerned, did not register any far-reaching achievements with respect to any of the three components. However, energetic beginnings have been made with respect to each of them. These include a better and clearer appreciation by the Arab governments of the imperativeness of developing the components of downstream operations, and the translation of this appreciation into plans, programmes, and projects. But most of these will come to fruition in the decade of the 1980s. It is worth noting, in this connection, that until the mid or late 1970s, considerably greater concern was felt and action taken with regard to upstream operations. This was partly a reflection of the system of priorities implicitly adopted, partly a reflection of the greater capability of the governments to deal with such aspects of upstream operations as finance, technology, or management, and partly a reflection of the lesser degree of control that the Arab oil producers could exercise with regard to downstream as against upstream operations. Although advances in the areas of technology, operation and management, and control still fall very short of the desired level, nevertheless they are felt to justify energetic action in downstream operations.

The second observation is that the present examination will not include a detailed survey of the steps taken in the development of downstream operations, although some summary information will be given. Such information will mainly aim at providing a factual background to the main purpose of the section, namely to identify the major issues involved in the development of the activities of gas treatment, refining, and petrochemical industry, and to point to the directions which such development should take in the years to come, if the hydrocarbon resources are to serve the Arab economies more fully and effectively.

Thirdly, again it must be admitted that the formulation of policies relating to downstream operations has not proceeded collectively by the Arab governments, whether within OPEC or OAPEC, but on a country-by-country basis. While this is also largely true of all upstream operations except crude pricing, nevertheless a great deal of consultation and discussion has been conducted by the community of Arab oil producers around upstream operations. Consequently, our examination of policies relating to downstream operations will in fact be based on individual government policies, but an attempt will be made to discern (or impute) the general trends and mainstream positions that underlie the country policies and activities. The merits of collective policy formulation and action will be pointed to, as a response to the economic logic of the optimal development of downstream operations.

The major concern with the development of downstream operations manifest in the years following the correction of crude prices in 1973/4 is in fact the summation, and later the integration, of a number of particular concerns with regard to crude exports, refining, gas processing or treatment, and the industrialisation of oil and gas. These individual concerns became forceful in all the Arab oil-exporting countries, though in varying degrees, and led to disparate conclusions and action policies. So far, the integration of individual policies has been taking place, where it is taking place within countries. But no policy integration among countries had occurred by the end of the 1970s or, indeed, by 1981 and 1982 (that is, prior to the completion of this study). But although no regional downstream policies (or OPEC policies, for that matter) exist, the individual country policies have a great deal in common, thanks to the similarity of problems, views, and analyses in a number of areas. It is these shared policies that will receive most of our attention here.

Several concerns have combined to promote the integration of all oil and gas-related activities within individual oil-exporting countries. Though these concerns have acquired great urgency and received concentrated attention and substantial resources only recently, most of them had been on the minds of policy and decision-makers for many years before the 1970s. A combination of factors had inhibited the entry by Arab governments into downstream operations on a substantial scale before the mid-1970s. We will attempt in the following discussion to identify the concerns in question, and the inhibiting factors, as we look at the present situation against its background.

*Refining*

The primary concern to consider within the area of refining is the expansion of refining activity in order for the Arab countries to be able to export an increasing proportion of refined products instead of remaining essentially confined to the export of crude oil. The immediate, multi-factor justification for such an expansion was the additional value added expected from each barrel of oil refined and exported, and the upgrading of the oil sector from one restricted to the production and export of a raw material, to one capable of processing and transforming this material, with all that goes with such upgrading by way of acquisition of new technological, managerial, and organisational skills; the construction of new (or expanded) industrial plants; the expansion of investment opportunities at home; the entry into new foreign trade relations; the earning of more foreign exchange; and, above all perhaps, the extension of control over the hydrocarbon sector.

It ought to be stated at this early point that refining activity had been undertaken for decades in the region, both in oil-producing countries, and in such non-oil countries as South Yemen, Egypt, Syria, Lebanon, and Jordan. Indeed, it had been the policy of foreign concessionary oil companies to build refineries near the resource, and in the early phases of the oil industry in the Middle East, most of the exports had been in refined form. It was in the late 1950s that the pattern was reversed, with an increasing number of new refineries becoming market-based — that is, built in Western Europe. The 'conventional wisdom' in the last two decades, as pressed by the companies, has been for the oil-producing countries to avoid a vast expansion in export-oriented refining activity. The argument behind the pressure is two-pronged: that there was by the 1970s considerable excess refining capacity in Western Europe which created an inappropriate climate for new capacity in the oil-producing countries, particularly in view of the higher cost of building new refineries in these countries;[63] and that, in any case, the oil-producing countries would be well-advised to adhere to the existing pattern of division of labour within which they would specialise in crude production and export and leave refining (except for the domestic market) to the industrial countries, with their greater capabilities in terms of infrastructure, technology, expertise, experience, and proximity to large consuming markets.

The reaction of Arab producers to this type of argument has several aspects. They acknowledge the fact that there is surplus refining capacity in Western Europe, but are equally aware of two other relevant facts: first, that part of this excess capacity arises because the configurations of modern refineries designed to take account of the changing structure of demand for products and the changing mix of light and heavy crudes, and of the changing needs of the petrochemical industry, are partly to blame; and secondly, that both the United States and Europe are still large net importers of refined products — the former to the tune of 62.7 million tons, and the latter 66.1 million tons in 1980.[64] In addition, the Third World will provide a large scope for Arab refined products once these become available for export on a substantial scale (which would mainly be in the mid-1980s).

In any case, the refining capacity of the Arab members of OPEC stood at 2,012 thousand barrels a day by the end of the 1970s;[65] this represented 9.4 per cent of their production for the same year, but only about 2.3 per cent of world refining capacity, as Table 3.4 shows. Surely, the Arab producers argue, as owners of crude who contribute about half of the crude oil in international trade they are entitled to a distinctly larger share in the export of refined products than the small fraction for which they accounted in 1979. The existing system of international division of labour, the producers add, should be no deterrent to the development of a considerably larger refining capacity: on the contrary, it should be an incentive, as the system needs drastic revision to take account of the aspirations of Third World regions, and of the new realities in the field of energy.

TABLE 3.4: **Capacity of Refineries in Existence, Under Construction, and in Planning in Arab Oil-exporting Countries (1000 b/d)**

|  | In existence 1979 | 1980 | Under construction | In planning | Total |
|---|---|---|---|---|---|
| Iraq | 323 | 335 | 200 | 200 | 735 |
| Kuwait | 644 | 644 | – | 106 | 750 |
| UAE | 15 | 15 | 180 | 140 | 335 |
| Qatar | 10 | 13 | 50 | – | 63 |
| Saudi Arabia | 760 | 765 | 1,185 | – | 1,950 |
| Libya | 138 | 138 | 220 | – | 358 |
| Algeria | 122 | 501 | – | – | 501 |
| Total | 2,012 | 2,411 | 1,835 | 446 | 4,692 |
| World Total | 79,600 | 81,300 | | | |
| Arab to World % | 2.5 | 2.96 | | | |

Source: OAPEC, *Secretary General's Seventh Annual Report 1400 H – 1980 A.D.*, Table 17, p. 70 and p. 65 for world capacity. (Based on official country records, *Oil and Gas Journal*, 29 December 1980, and *Arab Oil and Gas*, 1 and 16 September 1980.)

Furthermore, the high capital cost of construction of new refineries is to a not negligible extent artificial, in the sense that it goes well beyond the level justified by inflation in the industrial countries that undertake most of the design and construction in the Arab countries. (Indeed, there is evidence that price discrimination is practised against these countries by the suppliers of Western technology, whether in the form of hardware or software.) Finally, with respect to the drop in the profitability of refining as a result of the excess capacity currently in existence, Arab governments point to the advantage they enjoy in having the crude right at the point where refining capacity exists or is to be installed, and secondly, to the fact that they have a much longer time horizon incorporated in their evaluation of refining, compared with the profit motive. As to the oil producers' disadvantage in terms of inadequate infrastructure and insufficient expertise, this can be corrected in the relatively short term.

Taking all matters into account, the Arab governments were pressing for expanded refining capacity by the late 1970s,[66] but the implementation of their long-term plans seems to have since become more cautious. Yet, as Table 3.4 shows, planned new and/or expanded capacity involves 2,281 thousand barrels a day, which would bring the aggregate share of the Arab members of OPEC to 4,293 thousand barrels a day, equivalent to about 5.4 per cent of world's total capacity as it stood at the end of the 1970s. This more than doubling of Arab capacity by the mid or late 1980s would still account for 20 per cent of crude production by the Arab members of OPEC (on the basis of the level of production in 1979). If account were taken of domestic demand in the producing countries, estimated for our purposes at some 15 per cent of production by the mid-1980s, this would leave only a little over the equivalent of 1 million b/d of refined products for export, or at most about 18 per cent of the volume of refined products in international trade.[67] Again, this would be a far smaller proportion than the share of Arab crude in international trade. It is this empirical background that motivated OAPEC in December 1978 to set up a committee to explore the formulation of a common policy for the marketing of refined products to the countries of the European Economic Community and other industrial countries, in the hope that there would be no impediments to free access. This had been preceded by other manifestations of deep concern with the future of Arab refining industry, crowned by a special symposium on the subject held in Damascus in October 1975.[68] Another manifestation can be found in the studies published over the years in the Organisation's journal, *Oil and Arab Cooperation*.

However, this overall Arab concern to expand refining

activity has roused other concerns. We have referred to two of them in passing. They are, first, the lowered profitability of the industry in the main consuming countries owing to the excess capacity in existence; and, second, the need for a change in the configuration of refineries to take account both of the needs of the petrochemical industry and other oil and gas-related industries, and of the fact that high-sulphur heavier crudes are becoming more abundantly available than low-sulphur light crudes. Against this, there is growing demand for light and medium (at the expense of heavy) distillates. This makes it necessary to build in greater capability for advanced processing in order to adjust the stream of production to the needs of the market, and likewise to adjust for the emerging new profile of crudes coming in for refining. To these we must add the need to apply stricter specifications in order to combat pollution, and to encourage energy saving in the refining process. We need not go into the technical details of these issues.

Great stress must additionally be placed on certain economic and political problems that would arise in conjunction with the expansion of export-oriented refining capacity in the Arab countries. These include the difficulty of estimating future demand for products, in view of the speculativeness of the availability of crude in the future and the mix in which it will be available, the new products and product specifications that would emerge over time to accommodate future markets and anti-pollution regulations,[69] the role which research and technological development are likely to play in the development and promotion of certain types of advanced products, and the impact of restrictive regulations that the large industrial countries might well impose on the import of refined products.

But above all, we must consider one significant concern which the Arab countries will no doubt have to take into account in the present context. This is the likelihood that the industrial countries will feel restive and unaccommodating under the fear that there might be pressure by the oil producers to tie the export of refined products to that of crude, no matter how minimally unattractive the tie is made out to be. Furthermore, the industrial countries, with a preponderance of older, less advanced and less adapted refineries, will no doubt resent the advantages that the newer Arab refineries enjoy, being on the whole more advanced and better adapted to the emerging needs of the consuming market and the dynamic petrochemical industry.

The Arab producers cannot neglect this issue, even if it is far fetched, and the professional literature on the subject by Arab writers and officials suggests that they are seriously taking it into consideration. The issue goes beyond the tie-up between the export of crude and that of refined products, to reach that between crude and petrochemical products, once these are produced in enough quantities to press for export opportunities. Hartshorn speaks of the fear by industrial countries of excessive dependence on oil producers, and of the tie-up as a form of 'coercion' which the developed countries would be most reluctant to accept.[70] These are serious matters that deserve examination, but we will have occasion to consider them more analytically later on under the general area of downstream operations with its three components.

*Gas Treatment*

When we come to natural gas in the Arab countries, we find that the problem that has for decades beset the hydrocarbon industry is of a different quality from that in the case of refining or of petrochemicals. It is one of outright waste on a colossal scale, not just the insufficient development of potential as in the case of refining or petrochemicals. This dual aspect of the problem will become clearer once we provide some quantitative information in support of the statement.

As Table 3.5 will show, the known reserves of natural gas (both associated, that is, in company with crude oil, and non-associated) amounted to 11,136 billion cubic meters (or 11.1 trillion $m^3$) by the end of 1979 for the Arab members of

OPEC, which represented about 15.3 per cent of total world reserves as of the same date. Gas production during the year 1979 amounted to 166,859 million m$^3$, or 166.9 billion m$^3$, most of which (except in the case of Algeria) was associated gas. However, a little over half of the volume produced (89.5 billion m$^3$ or 53.6 per cent) was utilised; the remainder (77.4 billion m$^3$ or 46.4 per cent) was flared — that is, wasted. These figures indicate that the endowment of the Arab countries in terms of natural gas is much smaller, in relative terms, than their endowment in crude reserves, contrasted with world reserves: 15.3 versus 52.9 per cent respectively.

TABLE 3.5: **Natural Gas Reserves of the Arab Oil-exporting Countries, End 1979; Gas Production, Utilisation, and Flaring, 1979**

|  | Reserves billion m$^3$ | Production Million m$^3$ | Utilisation % | Flaring % |
|---|---|---|---|---|
| Iraq | 778.7 | 14,410 | 15.5 | 84.5 |
| Kuwait | 948.6 | 13,028 | 72.6 | 27.4 |
| UAE | 580.5 | 15,244$^a$ | 39.9 | 60.1 |
| Qatar | 1,699.0 | 6,580 | 67.5 | 32.5 |
| Saudi Arabia | 2,711.0 | 50,561 | 29.9 | 70.1 |
| Libya | 679.6 | 23,456 | 79.3 | 20.7 |
| Algeria | 3,737.9 | 43,580 | 77.0 | 23.0 |
| Total or per cent: | 11,135.3 | 166,859 | 53.6 | 46.4 |
| World total | 72,867.0 | 1,540,195 | c.90 | c.10 |
| Arab to world % | 15.3 | 10.8 |  |  |

*Note*: a. Abu Dhabi only.
*Source*: OAPEC, *Secretary General's Seventh Annual Report 1400 H – 1980 A.D.*, Table 15, pp. 55-6 and Table 16, pp. 62-3. (Based on official country records and different issues of *Oil and Gas Journal*.)

Whereas the proportion of Arab gas flared to the total produced is almost one half, it is only one-tenth for the world as a whole.[71] The excessive wastage in the case of Arab countries, which by the end of the 1970s amounted to about 60 million tons of oil equivalent a year, or 1.2 million b/d, has been inherited from the era of concessionary companies. Indeed, the rate of utilisation grew by more than two thirds

## New Policy Options in an Integrated Context

between 1973 and 1979, that is, under government control: from about 30 per cent to almost 54 per cent. Furthermore, it had been much lower in the 1950s and 1960s; its growth during the period preceding the take-over of control by the Arab governments is to be attributed almost wholly to the policy of these governments to use gas as a fuel, mainly for water-desalination and electricity-generation plants, but also for other industries, as well as for the production of fertilisers. In other words, it is not the companies that should get the credit for such utilisation.

The basic issue in this context is mainly one of political economy, though it is also partly technical. It is the reluctance of the companies to develop gas resources (whether through liquefaction or industrial use). In turn, this situation persisted under the concessionary regime because crude oil was priced very low and the opportunity cost of gas, which was considered a by-product, was estimated to be extremely low or nil. Furthermore, the companies were not motivated by the conservation of a national resource as the governments were and still are, nor by considerations of overall national development to which gas could make a tangible contribution in more than one way. Finally, the heavy capital cost of gas treatment plants further inhibited investment. Alternatively, the re-injection of gas into the oilfields was not called for in view of the high and easy yield of these oilfields. All this combined to enable the companies to justify (even to rationalise) gas flaring. The Arab governments on many occasions expressed their disagreement with the outlook and arguments of the companies, but the futility of their insistence that flaring should be considerably reduced, was part of their overall helplessness, which the take-over in the early 1970s came to remedy.

Since the take-over, Arab concern with the utilisation of gas has grown considerably, although not much collective deliberation and policy formulation has been undertaken by the governments. The most notable collective effort has come from OAPEC's Secretariat since 1973, through serious studies undertaken and seminars and conferences related, wholly or partly, to gas resources. These include several papers given at the five annual training programmes conducted so far on the basic aspects of the oil and gas industry; the relevant papers given at the First and Second Arab Energy Conferences held in 1979 and 1982 respectively;[72] articles and research papers published in the OAPEC journal, *Oil and Arab Cooperation*; studies and essays on the subject in collections published;[73] and a special symposium held in Algeria in 1980 on the utilisation of natural gas in the Arab world.[74] The combined impact of this considerable effort has been strong, as manifested in the active participation of government representatives in it, but more so in the seriousness with which the governments are examining and formulating their gas policies.

A review of developments in the 1970s permits some important generalisations in addition to the one related to the seriousness of gas flaring. The first is that gas liquefaction basically utilises associated gas, except in Algeria, and that this latter country pioneered LNG (liquid natural gas) export from non-associated gas. Its large projects involve export to the United States and Western Europe (the former through special tankers, the latter through tankers as well as a pipeline linking Algeria with Italy). Abu Dhabi was the first among the Gulf countries to export LNG from associated gas. Iraq and Kuwait have invested moderately in liquefaction and other gas treatment work but mainly for their fertiliser and petrochemical industries, with Kuwait using gas additionally for desalination and electricity-generation plants. Saudi Arabia and Libya have made no investments in LNG plants in the 1970s. Owing to the pre-eminence of Saudi Arabia in the oil industry as a whole, it is pertinent to explain that this country is not pursuing LNG export from associated gas, but intends instead to utilise gas in industry (based on methane and ethane fractions of associated gas), and to export large quantities of LPG (liquid petroleum gas) condensed from its associated gas. (See Table 3.6 for a summary of liquefaction projects in the oil countries.)

The second generalisation to make is that the slowness of

the development of gas plants is partly explainable by the very heavy investments involved (particularly with respect to non-associated gas) for processing and liquefaction, special liquid gas tankers, and re-gasification works at the importing end.[75] The large capital costs involved necessitate the entry by the producers and the small number of large importers into long-term contracts that would assure the amortisation of the heavy investment and the continuity of the trading relationship while amortisation is proceeding. However, such long-term contracts are particularly difficult to make owing to the impossibility of quoting one price for gas which will remain applicable during the fifteen or twenty years that the contracts would last. The inclusion of clauses that permit price adjustments during the operation of the contract is no easy way out, as both exporter and importer are hesitant to enter into substantial financial commitments without the ability, fairly clearly, to establish the size of the commitments in advance. The underlying problem here is that there is no market price for gas as there is for crude oil, and the transactions in gas are much fewer than in the case of crude. Each large, long-term contract has to be priced by itself, with contracts being few and far apart. This third generalisation regarding the pricing mechanism also partly explains the slow development of gas resources.

The fourth generalisation is probably more telling in the calculations and policies of the producers. This is that the price of gas, though it has been raised considerably in relative terms, still fails to reflect the real value of gas, and is still well below that of oil equivalent. This inhibits the investment and efforts needed to capture gas and to explore for non-associated gas reserves. Since gas is a substitute for oil for many uses, their respective prices must be close to each other. Admittedly, it is not easy to design a pricing formula for gas, any more than it is for oil; both tasks are complicated and rather inconclusive in results. But gas pricing presents some special difficulties connected with the calculation of transport costs, the differentiation between gases, the fact that the conversion to oil equivalent does not solve the problem of substitution in actual practice, the question of deciding what the 'right' netback would be and has to be, and the absence of a special market for gas to serve as a unifying mechanism and to provide a unifying indicator. As in the case of oil, sight must not be lost of the exhaustibility of gas (which means that 'economic rent' should be allowed for adequately, in addition to the allowance for the heavy investments involved). Furthermore, price should take into account the prices of near alternatives (such as fuel-oil for heating, naphtha as feedstock, gas oil for house use, and so on). Finally, the producers should capitalise on the technical and economic advantages of gas (quality, cheapness even if priced at par with oil, cleanliness), and on its versatility.[76]

TABLE 3.6: **Capacity of Gas Liquefaction Plants in Existence and Planned, 1980 (1,000 metric tons)**

|  | Status of Project | LNG | Ethane | LPG | NGL | Total |
|---|---|---|---|---|---|---|
| Iraq | In existence | – | – | 250 | – | 250 |
|  | Planned | – | – | 1,200 | – | 1,200 |
|  | Total | – | – | 1,450 | – | 1,450 |
| Kuwait | In existence | – | – | 6,009 | 2,192 | 8,201 |
| United Arab Emirates | In existence | 2,300 | – | 1,603 | 464 | 4,367 |
|  | Planned | – | – | 2,376 | 2,138 | 4,514 |
|  | Total | 2,300 | – | 3,979 | 2,602 | 8,881 |
| Qatar | In existence | – | – | 1,033 | 383 | 1,416 |
| Saudi Arabia | In existence | – | – | 6,500 | 3,000 | 9,500 |
|  | Planned | – | 2,685 | 8,617 | 2,904 | 14,206 |
|  | Total | – | 2,685 | 15,117 | 5,904 | 23,706 |
| Libya | In existence | 3,246 | – | – | – | 3,246 |
| Algeria | In existence | 20,200 | 343 | 600 | 3,100 | 24,243 |
|  | Planned | 10,100 | – | 5,030 | – | 15,130 |
|  | Total | 30,300 | 343 | 5,630 | 3,100 | 39,373 |
| Total | In existence | 25,746 | 343 | 15,995 | 9,139 | 51,223 |
|  | Planned | 10,100 | 2,685 | 17,223 | 5,042 | 35,050 |
| Grand total |  | 35,846 | 3,028 | 33,218 | 14,181 | 86,273 |

*Source*: OAPEC, *Secretary General's Seventh Annual Report 1400 H – 1980 A.D.*, Table 22, pp. 85-6.

## New Policy Options in an Integrated Context

The fact that the price of gas has risen somewhat, has made its treatment economically feasible. The process will go further, the closer the prices of oil and gas get. Nevertheless, the heavy initial investment in gas plants coupled with its relatively low price, means that the netback from export remains very low per unit of investment. The Arab producers, like other gas producers such as Mexico and Canada, and the major consumers as well, are now persuaded that gas prices will continue to rise until they reach or approximate those of crude. This process is inevitable in view of the growing conviction that the supply of energy (in all its forms) will be getting tighter in the medium term, and probably much more so in the long term, and will do so until alternative sources in commercial quantities and at tolerable prices are developed. It is generally thought that this will require at least one decade if not two. In the meantime, crude oil resources are coming under great pressure especially with the growing conservationist attitudes of the owners of these resources; and the pressure is stronger than the consumers' efforts in the large consuming countries to relieve it.

The impact of these various forces on supply and demand is directing the thinking of specialists and policy-makers to the consideration of gas as an 'energy source in its own right', but as bridging energy — one which can gradually and increasingly take over *some* of the 'responsibility' of crude oil and some of the pressure off it. This role of bridging or transition can only partly be played by gas, since known gas reserves cannot justify great hopes for the availability of gas for many years. The limitation arises from the relative modesty of gas reserves, compared with oil, and much more so with coal, as Table 3.7 shows. Nevertheless, this does not negate but only qualifies the significance of the role of gas.

This trend in thinking creates a dilemma for the Arab producers, and here the fifth generalisation comes in. The upward adjustment of the price of gas makes investment in its development attractive; the large oil revenues make the investment possible; the pressure by the large energy consumers makes gas exports assured for a considerable stretch of time. On the other hand, the Arab producers ask themselves three very pertinent questions: (1) Why should they use up their gas resources quickly, mostly as a fuel in the advanced industrial countries, in order to spare the gas reserves of these countries for subsequent use? (2) Though now in possession of adequate financing, why should they shoulder the burden of large investments to accommodate the importers of gas, and why should these not share in the fixed capital costs, since they are the final beneficiaries of the added volume of gas to be made available? (3) Is the sale of gas, essentially to be used as a fuel, the most defensible use, and should the producers not consider much more seriously the development of large petrochemical and fertiliser (and other gas-using) industries at home, where the utilisation of gas will be more rewarding financially and more promotive of overall industrialisation and development?

TABLE 3.7: **World Reserves of Fossil Fuels (Billion tons of oil equivalent)**

| Region | Natural gas | Oil | Coal | Total | Per cent |
|---|---|---|---|---|---|
| N. America | 7.9 | 5.1 | 124.9 | 137.9 | 22.3 |
| S. America | 2.6 | 4.1 | 1.7 | 8.4 | 1.4 |
| W. Europe | 4.0 | 3.3 | 32.0 | 39.3 | 6.4 |
| E. Europe | 27.0 | 11.1 | 203.0 | 241.1 | 38.9 |
| Middle East | 17.0 | 50.2 | 0.2 | 67.4 | 10.8 |
| Africa | 6.0 | 8.3 | 8.4 | 22.7 | 3.6 |
| China | 0.7 | 2.7 | 67.5 | 70.9 | 11.5 |
| Rest of the World | 3.4 | 2.8 | 25.8 | 32.0 | 5.1 |
| Total World | 68.6 | 87.6 | 463.5 | 619.7 | 100.0 |
| Per cent | 11.1 | 14.1 | 74.8 | 100.0 | |

*Source*: OAPEC, *Proceedings of the Symposium on the Ideal Utilisation of Natural Gas in the Arab World*, Algeria 29 June to 1 July 1980 (Kuwait, 1980, Arabic and English). Table 3, p. 42 in the Arabic section; originally from UNIDO, 'Petrochemical Industry to the Year 2000', Part One, pp. 38, 39.

While the generalisations made all raise issues and create concerns for the Arab producers, the last generalisation and the questions deriving from it hightlight the major concerns in the field of gas development for export. The Arab countries seem clearly to have reached a position on the issues suggested by our discussion. The ingredients of this position are the following: to concentrate attention on the utilisation of associated gas (except in Algeria with its much larger endowment of non-associated gas), and therefore to aim as a top priority at the reduction of gas flaring; to concentrate on the production of ammonia, methanol, and ethylene as feedstock for the petrochemical and fertiliser industries; to use gas in the aluminium, iron, steel, glass, and cement and clinker industries; and finally to re-inject gas into the oil wells as necessary, to maintain pressure and enhance oil recovery.

This position is different from that characterising the period preceding the 1970s. Then, gas did not claim a great deal of official attention, mainly because oil exports were more remunerative and gas in contrast was extremely unrewarding; the domestic consumption of gas was small; the world market, especially Europe, was self-sufficient in gas; liquefaction projects were too costly for the financial means of the Arab countries; and entry into the petrochemical industry was inhibited both by its complexity and the weight of discouraging advice given by the foreign oil companies. Algeria's calls for more attention to gas went largely unheeded until the drastic change in crude oil pricing in 1973/4 transformed the economics of gas utilisation favourably. But it was to be only early in 1980 that Canada and Mexico markedly raised the price of their gas exports to the United States, followed by the United Arab Emirates with its LNG exports to Japan, and by Algeria with its exports to the United States and France, and that the alignment of gas prices with oil prices became a pressing demand.[77]

By this time, another equally pressing demand was being strongly voiced and increasingly implemented by the Arab oil countries: that of directing the development of gas resources more towards the petrochemical industry than towards liquefaction and export.[78] This is reflected in their plans for the half-decade 1980-5, which emphasise the 'exploitation of existing mineral resources; recovery and exploitation of gas (especially associated gas); development of petroleum products processing from common refining to more sophisticated treatment of the oil barrel; establishment and/or expansion of basic petrochemical industries; general growth in sectors concerned with materials of construction; growth in the metallurgical sectors.'[79] The activities within this list which are directly and closely linked with oil and gas (namely, oil refineries, gas plants, petrochemical plants, chemical products, and rubber products) account for some $58 billion of investment as Table 3.8 shows. The projects relating to these activities are ones currently under study, planning, or execution, as well as ones already completed. The total number involved is 166, and only projects with an investment of $10 million or more are included.[80] Thus it seems that the future course of action is clear with regard to the avenues which the utilisation of gas should take, considering the thinking, the policies, and the pattern of investments in the Arab oil-exporting countries by the end of the 1970s.

*Petrochemical Industries*

This is the third component of downstream operations which are being examined in this part of the chapter. The term is used to comprise not only petrochemical industries narrowly defined, but also the fertiliser industry and other industries that use gas as a fuel or feedstock heavily, like iron and steel, aluminium, cement, and glass. Unless we want specifically to refer to these other users of refined petroleum products or of gas, we will employ the term 'petrochemical industry' as a generic term to include the whole group. It remains to be added that although the petrochemical industry is discussed in a section by itself, it forms part of the integrated view of downstream operations, as already indicated earlier.

The discussion to follow will hopefully demonstrate this interrelationship, but for the moment we will stress the point that while a petrochemical industry can physically be established far away from an extractive industry which produces crude oil, or from a refining industry which produces the refined products and a gas treatment industry which produces the gases needed as feedstock — while it is possible to do that, it makes much greater economic and technical sense to have all these industries located together. This is because of the linkages that are made both possible and convenient through proximity. The advantages translated into savings go a long way to compensate for the higher costs involved in the case of the Arab countries because of the insufficiently developed infrastructure, the insufficiently trained manpower, and the high capital cost of plant construction. Finally, as the fuel and feedstock needed constitute an inordinately large proportion of production (variable) costs, the location of the industry near oil and gas fields is particularly advantageous. These various advantages must be supplemented by others that accrue to an oil-producing country by entering the field of petrochemicals, as we shall see later.

We have had occasion to point to the absence of petrochemical industry in Arab countries before the 1970s, with the exception of the production of fertilisers in a few instances. The hesitation of the Arab countries to enter the field of petrochemicals narrowly defined can be understood, in view of three categories of reasons that combined to influence the position taken. The first was the objective unpreparedness and inability to enter the field with its complicated technology and large scale. This covered the inadequacy of scientific and technical skills — from design down to construction and operation — as well as the insufficiency of domestic financing, considering the very low prices of crude prevailing and the modesty of oil revenues earned. The second category was the awe felt in the face of the enormous marketing capabilities and tasks that had to be possessed and shouldered in competition with already established giant producers of petrochemical products — whether these were basic, intermediate, or final products. That these producers were transnational corporations (TNCs) with vast empires, made the prospect of 'crashing-in' particularly daunting. Indeed, it is still daunting today, even with the changed circumstances of the Arab countries. And the third category of reasons was the weight and influence of the advice which the oil countries got from the concessionary oil companies, which were themselves among the giants producing petrochemicals. This advice stood strongly against the establishment of petrochemical industry by the Arab countries. The negative attitude of the companies was based on three arguments: the unpreparedness of the countries technically and managerially, the harsh competitiveness of the market, and the wisdom of adherence to the existing pattern of international division of labour whereby the countries would concentrate on the extractive aspect of the oil industry, and leave downstream operation (except for refining for domestic markets) to the oil companies with their resources, market connections, expertise and experience.[81]

Not all the oil-producing countries were necessarily convinced by these arguments, but they were all duly inhibited. It was only in the latter part of the 1970s that their convictions changed, as well as their financial circumstances, but above all their formal power as the party now in control of the oil industry in its totality. This sea change with its far-reaching dimensions and implications made the countries determined to translate their new formal power into *effective* power. Hence their new refining, gas treatment, and oil-industrialisation policies and investments. (We have already mentioned the reservations of these countries with regard to gas treatment for export. While the entry into the areas of export refining and petrochemical industry constitutes an important departure from the kind of advice and manoeuvring exercised by the oil companies, the *abstinence* from massive entry into gas treatment for export is also a sharp departure, for the reasons given previously.)

A look at the distribution of investments among downstream operations as seen in Table 3.8 for the six countries covered, supplemented by a review of the investments made in Qatar (on the basis of information available in other

TABLE 3.8: Investments in Oil and Gas-related Projects in the Arab Oil-exporting Countries[a] ($ million[b])

| Branch of manufacturing[c] | Algeria | Iraq | Kuwait | Libya | Saudi Arabia | UAE | Total |
|---|---|---|---|---|---|---|---|
| Oil refineries | – | 845 | 1,626 | 2,389 | 9,655 | 1,075 | 15,590 |
| Gas plants | 4,126 | 1,265 | 1,110 | – | – | 2,633 | 9,134 |
| Petrochemical plants | 296 | 3,786 | 909 | 2,408 | 14,190 | – | 21,589 |
| Fertiliser plants | 1,175 | 2,560 | 130 | 1,695 | 370 | 233 | 6,163 |
| Chemical products | 542 | 438 | – | 1,473 | 56 | 236 | 2,745 |
| Rubber products | 550 | – | 147 | 32 | – | – | 729 |
| Total | 6,689 | 8,894 | 3,922 | 7,997 | 24,271 | 4,177 | 55,950 |

Notes: a. Information on Qatar was not available when the table was compiled. b. Only projects involving $10 million or more are included. c. Projects included have just been completed, or are under construction or study.
Source: ENI, *The Interdependence Model, Data Bank: Industrial, Infrastructural and Agricultural Projects in Arab Petroleum Exporting Countries*, (Rome, October 1981), pp. 39, 40.

sources) shows a clear pattern. This is that, oil refining apart, Algeria has directed more resources to gas plants than any other country, followed by UAE and Kuwait. As to the petrochemical industry, it has received very little financing in Algeria, none in UAE, but considerable financing in Kuwait (though less than gas plants). Iraq, Libya, Saudi Arabia, and Qatar have invested heavily in petrochemical plants, and Algeria, Iraq, and Libya have additionally directed important investments into the fertiliser industry. Again, oil refining apart, the allocation of investments shows that the seven countries are broadly divided into two categories: those with heavy investment in petrochemicals (above one billion dollars), namely Iraq, Libya, Qatar, and Saudi Arabia; and those with heavy investment in gas plants, namely Algeria, Iraq, Kuwait, and UAE. Iraq falls into both categories. Indeed, Iraq is the only country with large investments in the three industries: gas treatment, petrochemicals, and fertilisers. (Algeria and Libya also demonstrate strong interest in more than one of these three categories, with Algeria's interest being in gas plants and fertilisers, and Libya's in petrochemicals and fertilisers.) This pattern will continue unchanged for some years to come, owing to the long time lead and large investments required if substantial change is to take place. It will not therefore be risky to speak with finality about the pattern as holding for a few years to come even with the recent establishment in the Asian Arab oil countries of the Gulf Cooperation Council which is attempting to co-ordinate oil policies and even to undertake joint programmes and projects in this field.

It will be useful to present the pattern for petrochemicals all by themselves, since they occupy us in this section. This will be done in terms of physical production capacity, as it is divided among countries and among categories of products — basic, intermediate, and final — in Table 3.9. However, the distribution will not include a differentiation between projects in existence and others under construction or planning. It is sufficient for our purposes to say that the ones already in existence constitute a mere 10 per cent of the total for basic products (in Qatar, Iraq, and Algeria), 5 per cent for intermediate products (in Iraq and Algeria), and 13 per cent for final products (in Kuwait, Iraq, Libya, and Algeria). Therefore, the generalisation is warranted that the petrochemical industry in the Arab oil-exporting countries will only come to fruition by the mid-1980s. Furthermore, the plans for this industry are constrained by the fact that the range of intermediate products is still narrow, limited as it is to the production of four products: ethylene dicholoride, monovinylchloride, styrene, and ethylene glycol.

TABLE 3.9: **Organic Petrochemical Products, in Existence, Under Construction, or in the Planning Phase in the Oil-exporting Arab Countries, 1980 (1000 tons/year)**

|  | Basic products | Intermediate products | Final products |
|---|---|---|---|
| Iraq | 135 | 66 | 185 |
| Kuwait | 780 | 455 | 145 |
| United Arab Emirates | 450 | – | – |
| Qatar | 280 | – | 210 |
| Saudi Arabia | 3,637 | 1,269 | 1,204 |
| Libya | 1,200 | 290 | 298 |
| Algeria | 353 | 40 | 83 |
| OAPEC | – | – | 180[a] |
| Total | 6,835 | 2,120 | 2,305 |

*Note*: a. One of the joint companies established by OAPEC has plans to produce synthetic rubber, detergents, and carbon black, with the total capacity indicated.
*Source*: OAPEC, *Secretary General's Seventh Annual Report 1400 H – 1980 A.D.*, Table 20, pp. 78-9. (The source table presents detailed information on individual products.)

In global terms, total capacity is 6.835 million tons a year for basic products, 2.120 million tons for intermediate products, and 2.305 million tons for final products. Ethylene and methanol account for the largest part of capacity in the first category (89 per cent), with six other main products accounting for the balance. On the other hand, most emphasis is placed on plastics among final products, especially low-density and high-density polyethylene (which account for 61 and 21 per cent respectively of plastics). The other categories of final products are synthetic fibres, synthetic rubber, detergents and carbon black. But these together account only for 255,000 tons a year. Synthetic rubber, detergents and carbon black (with a combined production capacity of 180,000 tons) are planned to be produced by a joint Arab company being formed as a result of studies undertaken by the Arab Petroleum Investments Corporation, one of the companies established by OAPEC.[82] Finally, the table shows the pre-eminence of Saudi Arabia in the promotion of petrochemical industry in each of the three categories indicated, with respect to the range of products and the size of production capacity. (It also leads the list in refining, but it has made very minor investments in the other branches of oil and gas-related manufacturing which Table 3.8 detailed.)

TABLE 3.10: **Capacity of Existing Plants and Plants Under Construction or in the Planning Stage for the Production of Ammonia in Oil-exporting Countries, 1980 (1000 tons nitrogen/year)**

|  | Existing plants | Under construction | Under planning | Total |
|---|---|---|---|---|
| Iraq | 787 | – | 515 | 1,302 |
| Kuwait | 544 | 272 | – | 816 |
| United Arab Emirates | – | 272 | – | 272 |
| Qatar | 486 | – | – | 486 |
| Saudi Arabia | 148 | 272 | – | 420 |
| Libya | 272 | 272 | – | 544 |
| Algeria | 272 | 544 | 272 | 1,088 |
| Total | 2,509 | 1,632 | 787 | 4,928 |

Source: OAPEC, *Secretary General's Seventh Annual Report 1400 H – 1980 A.D.*, Table 21, p. 82 (quoting Mahmoud Izzat, 'Nitrogenous Fertilizers Projects Based on Natural Gas in the Arab Countries', in OAPEC, *Proceedings of the Symposium on the Ideal Utilization of Natural Gas in the Arab World* (Kuwait 1980), pp. 129-31.

Table 3.10 completes this presentation of quantitative information on the petrochemical industry broadly defined. It contains data on fertiliser plants – in existence, under construction, and in the planning phase. This branch of the industrialisation of the hydrocarbon sector is of special significance to all Third World countries, because of the essential role fertilisers play in increasing land productivity and agricultural production, in a large part of the world where returns per farmer and/or per unit of land are generally low, and where hunger and under-nourishment are serious social and economic ills and threaten to become increasingly oppressive.

The information presented in Tables 3.9 and 3.10 suggests a vast expansion in capacity in relative terms, the capacity under construction or planning being contrasted with very little capacity in existence during the 1970s. But this picture is misleading if not considered against world demand and capacity. Thus, the final petrochemical products envisaged by the mid-1980s would account for no more than 3 or 4 per cent of the demand for these products as it stood by the end of the 1970s in Western Europe and Japan. Indeed, as indicated earlier, there is a wide discrepancy between the Arab region as an exporter of crude and its position as a producer of refined products or of petrochemical products. According to the Petroleum Economist's *OPEC Oil Report*,[83] the whole membership of OPEC (Arab and non-Arab alike) supplied 84 per cent of the world's crude in international trade, but owned 6 per cent of world refining capacity and 3.2 per cent of the capacity of the world petrochemical industry, by the end of the 1970s.

If the fertiliser industry is considered by itself, this would reveal that current Arab production of ammonia is only some 4 per cent of world production; the addition of planned capacity would bring the proportion only to 6.5 per cent by 1985 – if all the projects under construction and/or planning are completed according to schedule. The status of the fertiliser industry is of special pertinence to our discussion, since ammonia, which is the base for 97 per cent of nitrogenous fertilisers in the world (100 per cent for the Arab countries) is in abundance in the Arab region. Some 90 per cent of Arab ammonia comes from natural gas (70 per cent for the world), and only 3 per cent of natural gas produced in the Arab countries is used in the production of the ammonia in use. The fact that fertilisers produced in the Arab oil-exporting countries depend totally on ammonia has vast economic significance. It means that 65 per cent of the cost of production is accounted for by the feedstock, which is available locally.[84]

The discussion so far conducted has focused on each of the main activities within downstream operations separately, but it has on occasion referred to the interdependence among them. The point needs no further elaboration, as the case for close integration of refining, gas treatment, and petrochemical (and other related) industries is clear. Suffice it to say that refining produces some products which, together with certain gas components which gas treatment provides, go into the petrochemical industry as feedstock — to say nothing of the use of gas as fuel in the industry on the one hand and in refining on the other. (Gas liquefaction and export stands on its own as an important downstream activity, which can be undertaken without strong linkage with refining and petrochemical industry. Its individual significance cannot be overlooked, particularly as one source of bridging energy to help bring about a smooth transition from the age of the predominance of oil, to the age of the predominance of other sources of energy. But, as indicated earlier in this section, the Arab countries are rightly cautious in the export of gas, for good economic and strategic reasons.[85])

The case for heavy involvement in downstream operations is very strong. Although certain aspects of this case have been touched upon, it ought to be presented in more complete form. The Arab producers argue cogently for the involvement on several grounds; these will now be listed with some discussion:

1. Transition from the stage of crude oil export to that of oil and gas industrialisation, with all that that involves in technical and economic benefits, as the following points will show. In general terms, the transition will mean a greater diversification of the Arab oil economies. Thus, the present extremely heavy dependence on crude oil, which now accounts for 95-8 per cent of total exports, will be reduced and spread more evenly over other activities and exports. Although the new activities will still be related to oil and gas, they will introduce new dimensions to the economies and bring them to a more advanced stage of development.

2. Better integration of the hydrocarbon sector with the national economies. This point is related to diversification. Specifically, downstream operations will create some backward but many forward linkages, by promoting new industries and activities. The linkage will be direct in the case of the use of refined products and gas both as fuel and as feedstock, but also indirect in the activity to be generated in some other sectors and industries which will provide the downstream industries with, or receive from it, inputs and/or products.

3. The provision of vastly expanded and new opportunities for investment. The significance of this advantage is particularly telling in view of the somewhat limited opportunities in existence and the pressure therefore to export surplus financial resources to the industrial countries for placement or investment, with all that such export involves in terms of dependency and insecurity, and loss of real purchasing power arising from inflation and currency depreciation.

4. Greater balance in the world position of the oil countries. The background to this point is that there is presently an enormous imbalance between the place of the oil countries as exporters of crude, and that envisaged for them as producers/exporters of refined products, petrochemical products, and fertilisers. This imbalance has serious economic implications which spill over into the

area of power politics, and which restrict the ability to take relatively more independent politico-economic decisions and pursue a more self-reliant development.

5. Capitalisation on the natural advantage deriving from the existence of a richly-endowed resource base right where the downstream operations are conducted. This advantage goes a long way towards redressing the disadvantage of having to pay excessively for the capital costs involved and thus to bear an inordinately large rate of amortisation.

6. Acceleration of the acquisition of technological capability at home. This would be truer, the greater the involvement in the development of downstream operations: in the design, specification, installation, and operation of the plants and facilities called for by the various industries and activities, and in their many branches and ramifications. The skills which Arab manpower will learn will be of immense value in the industry itself, and also in the economy at large. The addition to the reservoir of skills will help upgrade the general capability of manpower through the development of associated institutions and activities (including science promotion, research and experimentation, training, and the many other professional services which will have to be developed parallel to the development of the hydrocarbon sector).

7. The increase in value added in every unit of gas or oil produced. This is particularly true of the petrochemical industry at the present juncture, with the netback of refining activities being rather depressed because of the excess capacity in Europe. But the advantage under discussion will be the greater, the sooner the Arab countries develop their downstream operations, and the sooner they succeed in penetrating world markets, especially in the Third World. To achieve such penetration, they have to demonstrate a fair degree of competitiveness in those markets, and to succeed in elaborating arrangements of joint ventures or composite transactions of import versus export.

*New Policy Options in an Integrated Context*

8. At the regional level, the promotion of closer economic ties and complementarity. The oil sector, even as it stands, has had a complementarity-promoting influence, but it has also had a certain divisive effect.[86] But the development of downstream operations will force the oil-exporting countries to seek the benefit of a much larger regional market for their downstream industries. In this context, aid-flows to the oil-importing Arab countries (as, indeed, to other developing oil-importing countries) will have a multiplier effect on trade, thanks to the availability at that point of more goods that the aid donors would have to sell, and to the import-generating effect of such sales.

9. Finally, enhancement of the position of the Arab oil-exporting countries (and, by extension, of the Arab region as a whole) within the Third World, thanks to the strength that the development of downstream operations and accelerated industrialisation would give to the region. This strength would be reflected in the Arab efforts along with other Third World regions to redress the injustices inherent in the existing International Economic Order and to build a new order. Likewise, the strength, if properly capitalised on, could also be put in the service of Arab national causes.

It might be argued that Arab oil policies have not been formulated on the basis of the case for downstream operations as just presented, or that no oil policy-maker has externalised all the arguments for the case. This may be true. But it is equally true that one or more of these arguments have been variously presented, by one policy-maker or another, even if not systematically. It is the analyst's duty to identify and integrate the arguments into one case, as this discussion has attempted to do. In the second place, it might be argued that the Arab oil-exporters, even when fully aware of some of the advantages listed, have not always consistently pursued them, or acted in a way that would maximise these advantages, or at least produce that mix of advantages acceptable to them from their own standpoint. This, too, may be true. But again it ought to be remembered that one decade — the 1970s, in the present context — is very short in the life of nations, and the learning process is a long and demanding one. The Arab countries will have to have much more time for refining their capability to formulate policies, and strengthening their hold on the mechanisms that serve the implementation of the policies. On the whole, in spite of the exceptions, it can be said that the Arab oil-exporting governments have gone through much of their apprenticeship speedily and creditably. That more clearly-conceived and determined action is needed to formulate and pursue the right promotive policies does not negate the value of what has already been achieved.

Finally, it might be argued that downstream operations have not been developed within a collective Arab framework, but individually by each of the countries concerned, and that co-operation and co-ordination are even less in evidence in the present area than in that of upstream operations. This is true. But there is evidence that the economic logic of the situation presses increasingly on the participants to act in concert — at least to deliberate together and to co-ordinate policies and actions. It will take some time for individual policy-makers to accept the notion, and then the practice, of jointly preparing the framework for policy-formulation, or at least to co-ordinate policies and avoid duplication and/or contradiction in oil policies. But the benefits accruable from such a course of co-operation and co-ordination are becoming more clearly demonstrable.

OAPEC has for years been contributing considerably to the cause of co-ordination, as are some analysts and writers.[87] In addition, most regional meetings concerned with hydrocarbons have recommended joint Arab action. Such recommendations have centred around the effective utilisation of associated gas for the production and marketing of fertilisers and petrochemical products. They have also urged collective action leading to more Arab consultation, co-ordination of planning, implementation of projects, marketing and credit facilities, technical back-up services, exchange of information, and development of skills (in the areas of design, construction, operation, maintenance, research, and product adaptation). Finally, the recommendations have urged collective action for the promotion of skills in dealing and negotiating with other regions, and for the attraction of investments into specialised regional institutions for marketing, engineering, training, and research aiming at the development of downstream industries.[88]

Some tangible results of these pressures for co-operation have appeared in the form of joint companies or projects sponsored by OAPEC in some areas both of upstream and downstream operations. But the cumulative effect of economic necessity and cogent analysis is yet to make itself felt in collective policies and efforts. The advantages of joint action — pooling and better allocation of resources on the basis of rational division of labour, an unwasteful approach to production, a concerted approach to regional and international markets, the drive to cut a place in the world export refining and petrochemical and fertiliser industries — will probably lead to the speeding up of co-ordination and co-operation.

Yet this rather rosy picture of the advantages of the development of export downstream industries, the optimistic outlook for such industries reflected in the last paragraphs, and the expectation of Arab collective policies must be tempered with a number of qualifications. These will come in the form of issues and concerns which should be faced and solved if the rosiness and the optimism are to be justified. An attempt will now be made to present the major issues involved in separate groupings.

## *The Economics of Downstream Operations*

The high capital cost involved in these operations is a problem to contend with. But the cost is particularly high for gas treatment, as indicated earlier on when gas was being discussed, less so for refining and petrochemical industries. As the direction of the analysis has been to reduce the emphasis on gas liquefaction for export, the problem will be contained. Furthermore, what is at issue is not the absolute size of the investments involved, since financing is available, but the fact that capital costs for the same plant are higher in the Arab than in the Western countries. The answer is therefore that the differential in costs should, and could be absorbable, as Kawari clearly argues in a paper already referred to. A study conducted by the United Nations Industrial Development Organisation, UNIDO, in co-operation with the Gulf Organisation for Industrial Consulting and OAPEC,[89] sums up its findings on the subject by saying that 'feedstocks and energy account for a significant proportion of total manufacturing costs. There is therefore a good prospect that such industries established in developing countries can be internationally competitive. The main requirements are that investment costs *do not greatly exceed those in the industrialized countries and that gas is made available at favourable prices.*' (Emphasis added.) To the cheapness of resources must be added the saving in transport costs involved; these costs will be lower if manufactured products are shipped than if the basic and intermediate products and the gas are to be shipped.

Another economic aspect to consider is refining capacity in the industrialised countries. As indicated earlier, in spite of there being some idle capacity in Western Europe, there is no parallel phenomenon in the United States, while both still import substantial quantities of refined products. And anyway, the planned expansion in Arab refining will still leave total Arab capacity a small fraction of world capacity, again as indicated earlier. Furthermore, much of the excess capacity in Western Europe is the product of the changing configuration of refineries and the ageing of many European refineries.

The situation is different with respect to petrochemical and fertiliser industries, where there is no serious problem of excess capacity. And new Arab production capacity will not make the Arab share of world industry any larger than 5 or

6 per cent of the total. Finally, there is wide scope for fertilisers and petrochemicals in the Third World at large. The entry of petrochemical products into many industries (like construction, transport, household equipment, and so on) is at an early stage, and much wider scope lies ahead.

Marketing, though tied to production capacity, deserves singling out by itself as an issue of importance. The critics of the development of petrochemical industries and refining in the Arab world emphasise two points: first, that they do not object to the establishment of such industries for domestic markets, but that it would be reckless to believe that the newly-industrialising countries could compete in the world markets with the products of the industrially-advanced countries; secondly, that if the newly-industrialising countries were to restrict their ambition to their own markets it would not be economical to produce a number of products which have to be produced on a large scale for technical/economic reasons. This kind of (rather contradictory) arguing calls for a number of reactions.

First, that ability or failure to compete is a risk the new producers will have to take, but that they should be competed with fairly. Secondly, if competition is fair, then what determines the size of the market accessible to the newly-industrialising countries is the prices charged and the quality assured. If the new entrants wish to take the challenge, they should be able to do so at their own risk. Thirdly, the planned capacity of the Arab countries will still be a very small proportion of the world total, and it could not be feared to constitute an unbalancing factor in world trade and the world economy. Indeed, the 5 or 6 per cent of world capacity expected to come on stream over the first half of the 1980s can only be of marginal impact on the established industries. And, in any case, the Arab oil exporters have a right to insist on 'going industrial'. Indeed, it could well be argued that those Western countries that exploited the economies of the Arab region for many generations − in oil and non-oil sectors alike − have a duty to help this region go industrial, especially in that it has proved very accommodating with respect to the export of crude and the 'recycling of funds', and to the limitation of its industrialisation ambitions.

The crucial question remains: will the oil-exporting countries be able to compete on the grounds of price and quality? According to general studies already referred to, and to the feasibility studies made for the projects under construction, the answer is affirmative on both counts. But here again, either the new entrants prove competitive, in which case the criticism based on the grounds of non-competitiveness collapses; or they prove non-competitive and fail to enter world markets, in which case the industries of the industrial countries have nothing to be concerned about.

## *The Infrastructure of Downstream Operations*

The first issue here is the provision of the physical infrastructure that downstream industries call for: communications, means of transport, pipelines, factories, and all the supportive physical facilities needed for the setting up, the operation, and the success and growth of new, complicated industries — ones which are market-oriented though resource-based, but above all technology and capital-intensive. Though the provision of such an infrastructure is costly in a region starting from a very low base, it can be and is being done — even if at a very high cost. It will not be allowed to act as a deterrent, judging by the course of development in all the countries concerned.

The technological infrastructure — the development of basic and applied science, research, and the training of manpower — constitutes in this writer's view the real handicap, not finance, prices, or markets. And this is precisely the area where the Arab countries have made least progress.

It is necessary to stress the danger and the social and economic (and, indeed, the political) costliness of the course that has been steered in most of the cases: that of the 'importation' of capital goods and expertise, in the belief that this is a short-cut to the true acquisition of scientific and technological capability. Evidently, the speed with which most of the Arab countries have desired to build up their industry has led them to the belief (perhaps the self-delusion?) that *there is* a short-cut to such acquisition, without realising that the short-cut is indeed the long road. This issue will have to be faced courageously and wisely if downstream industries — in fact, the whole drive for industrialisation and development — are not to remain accretions on the body of society and economy and not an organic part of its structure and capability.

Finally in this connection, training — though essential — will prove insufficient to mobilise the national manpower to participate actively and massively enough in downstream industries. This manpower will have to become much more strongly motivated and committed, and to that end social values have to become conducive to such motivation and commitment. Heavy dependence on expatriate manpower, which is the answer so far provided for the problem, will have to be reduced speedily and effectively.

## The Politics of Downstream Operations

The industrial countries of the West feel, or claim to feel, concern over the development of downstream industries in Arab (and non-Arab) OPEC countries. This concern has three aspects. The first is that if they are made to make room in their markets at home (or their established markets abroad) for Arab refined products and petrochemicals and fertilisers, through some linkage between the export of crude and that of products and petrochemicals, then they would submit to unacceptable dependence on the Arab (and other OPEC) countries. The second aspect is the fear of loss of advantages which Western refining and petrochemical industries currently enjoy on a world-wide scale, with all that that means in the extension of the scope of idle capacity, reduced trade, foregone profits, and fall in employment. The last aspect is the concern that OPEC members are spurred to develop downstream industries essentially by political motives.[90]

These are sources of concern that cannot be lightly waived

aside, even if in part they were to prove the product of auto-suggestion or political conditioning, or of self-inflicted worries. Whatever the source of fear of excessive dependence on OPEC countries, it remains vastly exaggerated. To tie the sale of refined products and petrochemicals to the Western countries to the sale of crude, is not possible. OPEC countries do not have enough leverage to do that, even if they were to opt for it. There are economic and political realities that deprive them of such leverage, and the Western countries have, and the Arab countries know that the Western countries know that they have, counter-leverage in terms of exportable-but-withholdable technology, food supplies and arms supplies. To this must be added the hold which the Western countries have on several Arab oil exporters by being the repository of their surplus funds, as well as the political hold the former have thanks to the alignments which the latter countries have chosen to adopt with the West. Finally, the cynic might ask: Is it not right that the Western countries should feel some limited economic dependency on the Arab countries, after all those decades of utter dependency — indeed of subordination to the West — suffered by the Arab countries, militarily, politico-strategically, economically, and culturally?

In answer to the second source of concern, namely the loss of advantage, markets, profits, and employment, it can be repeated that the scale of downstream industrialisation in the Arab countries is so small in contrast to that in the West that the harmful impact — even if not redressed — would remain marginal. However, the argument can be countered by the force of analytical logic and historical experience: that the expansion of trade from the Arab region to the industrial countries would enable the former to import more from the latter. Another aspect of recycling would thus be set in motion. Lastly, probably much of the market potentially to be gained by the Arab future exporters would be in Third World countries, which are nobody's private preserve.

Finally, the charge that the development of downstream industries would be largely motivated by political considerations, and the insinuation that it would not be motivated and justified by economic factors, can be easily cleared. The objective critic need only consider the advantages (listed earlier) of such development to see that the case for downstream industries is essentially economic and technological, though, admittedly, the emergence and success of these industries would satisfy national pride as an added bonus.

Assuming the gradual but firm predominance of economic common sense and fairness in the structure of relations between the Western industrial countries and the Arab oil-exporting countries, spurred by the power of mutual interests and interdependence, the issue becomes: Will the Arab countries be able to make the industrial countries extend technical help to them in the building of downstream industries, at terms that are fair financially and conducive to the acquisition by the Arabs of technological capability?

This is a serious issue, and it can be solved only if the assumption is based on a definite change of mind and heart in the industrial countries, thanks to which they would see the oil-exporting countries neither as a farm to exploit nor as a rival to fear, but as a co-producer in a world market which is extensive and growing, which provides room both for the industrial and the developing worlds, and in which the prosperity of the one helps the prosperity of the other provided no exclusivity is built into the international economic order. In such a world, the transnational corporations (TNCs) will have to submit to internationally-agreed rules of conduct in a manner which allays the fears of the developing countries of the hegemony and exploitativeness of the TNCs, and makes co-operation with them at least as beneficial to the developing countries as it is to the TNCs.

Another issue arises within the context of the politics of downstream industries, but this one is regional in nature. It relates to the question: How far will the oil-exporting countries seek to co-ordinate their policies and co-operate in the allocation, design, and construction of downstream industries; how far will they go in undertaking joint research and training; how far will they seek vertical and horizontal integration of their oil industries; how far will they go in attempting to turn the whole Arab region into one market for their own benefit and that of the Arab oil-importing countries; and how far will they co-operate among themselves in their dealings with the other regions of the world with respect to technology, award of construction contracts, dealings with the TNCs, pricing of products, and marketing in foreign markets?

As indicated earlier, there are economic and political pressures building up and pushing in the direction of co-ordination and co-operation, particularly among the oil-exporting countries in the Gulf area. But these have yet to be transformed into concrete collective programmes of substance – ones that truly reflect a full appreciation of the economic (and political) logic of co-operation, in a region with so many integrative ties, and in a world where large associations of countries can enjoy much more room for beneficial economic action than individual countries. An added pressure is the immense power of the TNCs, which cannot be counter-balanced except through large associations of countries acting together. Without such 'countervailing power', individual Arab oil exporters cannot carve a market for themselves in regions and countries which the TNCs consider their own preserve. And these seem to comprise virtually the whole world outside of the centrally-planned economies.

Finally, the case for co-ordination and co-operation among oil countries (and in fact all the Arab countries) is all the more urgent owing to the consolidation of separatist tendencies and structures. The slower co-ordination and co-operation proceed, the more difficult it becomes to dislodge separatist vested interests in favour of collective action.

This long chapter is coming to a close. Its length is justified by the need to treat upstream and downstream operations within the confines of one chapter, in consistency with the position reflected in the title ('New Policy Options in an Integrated Context'). The attempt has not been made at every linkage point between policies to indicate where and how closely the integrated aspect lies, yet enough has been said, it is hoped, to suggest that all the policies and policy options identified and examined fall within one integrated context. Essentially, this is the logic and necessity of treating the hydrocarbon sector as one entity, though it consists of separate, identifiable groups of operations. The development of this sector can be optimised only if all these groups are developed within general (sectoral) guidelines and objectives. And, in turn, the sector can best be developed only if this is achieved within the broader guidelines and objectives of the development of the whole economies of which oil forms but one sector. At this point, therefore, it is necessary to pose the question: How does the oil sector fit into the overall development of the oil countries, individually and as parts of the wider Arab region of which they form an integral part? It is to this question that the next chapter addresses itself.

# 4

## OIL AS ENGINE OF DEVELOPMENT

### The Financial Implications of the Policy of Control

As indicated in Chapter 2, the policy of control adopted in the 1970s involved the take-over by the governments of OPEC countries of the power to formulate and implement policies relating to the oil industry in all its phases, from prospecting and exploration all the way to the operation furthest downstream. Three policy areas within this wide range had vast financial implications for OPEC countries – Arab and non-Arab alike. These were determination of the volume of production and exports, determination of the export price of oil, and determination of the terms under which foreign companies (wherever they were still in special contract relations with the governments) were to operate, particularly with reference to the rebates or commissions (or the differential between their payments to government and their selling price, if they still were under participation arrangements) which they were to enjoy for every barrel of oil lifted by them.

Thanks to the application of the policy of control, the governments were able to introduce substantial corrections in the price of oil. These were mostly to permit the price to rise, in compensation for the decades during which it had remained artificially depressed during the concessionary regime (as we showed in Chapter 3), and to compensate for the steep inflation of the 1970s. That the price of oil had been extremely depressed can be evidenced by comparing it to the prices of close substitute sources of energy, and by the fact that today,

## Oil as Engine of Development

about one decade after the initiation of substantial price correction in the autumn and winter of 1973, the demand for oil is still greater than for any other source of energy.[1] Likewise, the failure of price correction for inflation to catch up with the rate of inflation can be seen from a comparison between the course of *current* prices over the years, and that of *real* prices (adjusted for inflation) over the same years. This comparison is made in Table 4.1. The table shows, first, the effect of inflation on the course of prices, and secondly, the effect of fluctuations in the value of the US dollar in terms of major OECD currencies during the same period. Taking account of dollar fluctuations is necessary in view of the fact that the price of crude is quoted in dollars, and payment for crude imports is also made in dollars. (This is apart from the fact that placements/investments in the US dollar or in the United States are the largest single component in the foreign reserves or surpluses of the oil-exporting countries.)

As the table reveals, the real price of oil has been lower for every year shown than the nominal (current) price charged, and the discrepancy has been larger for every year (except 1976) when account is taken of dollar fluctuation. But what is more serious, the real price of crude (after accounting for inflation) was actually lower for 1975, 1976, 1977 and 1978 than it had been in 1974. (The same applies for the real price after deflation for dollar fluctuation for all the years listed except 1976.) Thus, the real price for 1979 was 19 per cent higher than for 1974 (or only about 6 per cent if the real price plus dollar fluctuation is considered). Furthermore, the real price of oil in 1979 was only about 64 per cent of the nominal official price for that year (57 per cent if dollar fluctuation is accounted for as well as inflation). What makes these comparisons all the more telling with respect to the real modesty of price adjustments for inflation and dollar fluctuation, is that we are comparing 1979 prices with the base year, 1974, when 1979 witnessed sharp increases. Had it not been for these increases, the gap between current and real prices would have been much wider still than the table shows it to be.

It is against this background that the rise in the volume of

TABLE 4.1: Official Prices of One Barrel of Arabian Light Crude: (A) Deflated for Inflation Rates in OECD Countries; (B) Deflated for Inflation Rates in OECD Countries and for Dollar Fluctuation vis-à-vis Major Currencies (US dollars/barrel)

| Base year | 1974 | 1975 | 1976 | 1977 | 1978 | 1979 | 1980[a] |
|---|---|---|---|---|---|---|---|
| Official Price | 9.56 | 10.46 | 11.51 | 12.40 | 12.70 | 17.84 | 28.67 |
| (A) Deflated for inflation | 9.56 | 9.39 | 9.51 | 9.43 | 8.95 | 11.37 | 16.24 |
| (B) Deflated for inflation plus dollar fluctuation[b] | 9.56 | 8.99 | 9.66 | 9.03 | 8.12 | 10.11 | 14.35 |

*Notes:* a. Source states that the inflation rate for December 1980 was considered to be equal to that for November 1980, as the December rate was not available when the table was constructed. The same applies to dollar fluctuation.
b. Calculated on the assumption that the imports of oil-exporting countries are divided in the ratio of 20 per cent from the United States and 80 per cent from the five countries against whose currencies the dollar fluctuated.
*Source*: OAPEC, *Secretary General's Seventh Annual Report 1400H - 1980 A.D.* (Kuwait, 1981, Arabic), Table 27, p. 95 and Table 28, p. 97. (Original sources: *Petroleum Intelligence Weekly*, various issues, for official prices of crude; OECD, *Economic Outlook*, for rates of inflation, and IMF *International Financial Statistics*, for the dollar rates.

oil revenues between 1974 and 1979 must be understood and assessed. (It will be recalled that the years 1974-9 represent the period which witnessed price corrections and adjustments in the 1970s.) The growth of revenues which is recorded in Table 4.2 should be qualified in three ways. First, the data in the table refer to current dollars. Thus, while revenues rose by 162 per cent between the two terminal years, they only rose by 55 per cent (49 per cent if dollar fluctuation is accounted for), in real terms after deflation. Secondly, the revenues shown arise from the export of crude and of refined products. Although the latter are modest, they are not negligible. Finally, the revenues reflect both the level of prices and the volume of production. The latter rose by 19 per cent between 1974 and 1979, leaving that part of the rise accounted

*Oil as Engine of Development*

TABLE 4.2: **Oil Production and Revenues of the Seven Arab Oil-exporting Countries, 1961-79 (Production 1,000 b/d; Revenues $ million/year)**

| Year | Production | Revenues | Annual average revenue/b $ |
|---|---|---|---|
| 1960 | 4,333.2 | n.a. | – |
| 1961 | 4,748.5 | 1,190.2 | 0.69 |
| 1962 | 5,429.5 | 1,299.7 | 0.66 |
| 1963 | 6,230.0 | 1,511.3 | 0.66 |
| 1964 | 7,275.0 | 1,789.4 | 0.67 |
| 1965 | 8,170.5 | 2,159.8 | 0.72 |
| 1966 | 9.349.2 | 2,663.1 | 0.78 |
| 1967 | 9,804.8 | 3,021.8 | 0.84 |
| 1968 | 11,502.1 | 3,644.9 | 0.86 |
| 1969 | 12,549.6 | 3,937.6 | 0.86 |
| Total 1960-9 | 79,392.4 | 21,217.8 | |
| Average 1960-9 | 7,939.2 | 2,357.6 | 0.77 |
| 1970 | 13,826.4 | 4,534.3 | 0.90 |
| 1971 | 14,691.1 | 6,305.0 | 1.18 |
| 1972 | 15,751.6 | 7,705.4 | 1.34 |
| 1973 | 18,009.8 | 12,491.4 | 1.90 |
| 1974 | 17,723.3 | 51,499.3 | 7.96 |
| 1975 | 15,985.1 | 55,616.6 | 9.53 |
| 1976 | 18,579.4 | 66,415.3 | 9.79 |
| 1977 | 19,176.1 | 77,812.8 | 11.12 |
| 1978 | 18,455.4 | 73,522.4 | 10.91 |
| 1979 | 21,094.1 | 134,916.0 | 17.52 |
| 1980 | 19,233.8 | 204,244.0 | 29.01 |
| Total 1970-9 | 173,292.3 | 490,818.5 | |
| Average 1970-9 | 17,329.2 | 49,081.8 | 7.76 |

*Notes*: a. Data for revenues in the 1960s relate to the 9 years 1960-9. b. Average for revenues in the 1960s refers to the 9 years 1960-9. c. Revenues relate to earnings from the export of crude and refined products.

*Sources*: For production and revenues during the period 1961-80 *Middle East Economic Survey*, vol. XXV, no. 1, 19 October 1981, reproduced from OPEC's *Annual Statistical Bulletin 1980*. Production for 1960, from Ian Seymour, *OPEC: Instrument of Change* (Macmillan, London, 1980), Statistical Appendix. (For revenues 1970-80, also see Secretariat-General, League of Arab States, Arab Fund for Economic and Social Development, and Arab Monetary Fund, *The Consolidated Arab Economic Report, 1981*, Table 6/10, p. 302. (The latter source agrees with MEES only for 1979 and 1980, as it includes earnings from the export of refined products as well as for earlier years.)

for by price increases 36 and 30 per cent respectively (with or without dollar fluctuation). This is much more modest than the data on revenues in Table 4.2 suggest. Furthermore, the year 1979 was exceptional in that the level of production in it was the highest ever attained since oil production began in the Arab region; it also exceeds the level for 1980 or 1981.

Oil revenues of the Arab members of OPEC represented about 64.4 per cent of their aggregated gross domestic product, GDP, for 1979, estimated for that year at $209.6 billion.[2] But the significance of oil export revenues is even greater if related to total exports, total government budget revenues, or total earnings of foreign exchange. As indicated earlier, at the extreme they represent some 95-8 per cent of total exports, and almost as much of foreign exchange earnings. Finally, although it is difficult to assess the significance of oil revenues as a component of total budget revenues for each of the countries under consideration, we find they represent 66 per cent for Iraq and 78 per cent for Algeria – the two countries where the proportion is lowest.[3]

According to Table 4.2, aggregate oil revenues for the seven countries under examination totalled about $491 billion for the decade of the 1970s, but those of 1979 alone accounted for 27 per cent of the aggregate for the decade. (The revenues in the 1960s totalled some $22 billion only.) And even if we take the five years 1974-8, we find that revenues averaged only $68.6 billion a year, which is less than half the revenues of 1979, which was the year that marked the frontier of relatively large inflows. This comparison is made in order to show that the oil-exporting countries have not been floating in opulence since, and as a result of, the correction of oil prices and the expansion of production. Indeed, the aggregate GDP for all 21 Arab countries – oil-producing and non-producing alike – was $282.3 billion for 1979, while Italy's GDP alone for that year was $306 billion – that is, higher by 8.4 per cent than aggregate Arab GDP. The aggregate population of the Arab region was about 162 million in 1979, as against 57 million for Italy.[4] Furthermore, Italy's GDP derives from current production, not from the sale of a depleting, non-renewable resource as in the case of the Arab countries where oil revenues represent about 50 per cent of aggregate GDP. If we consider the Arab oil-exporting countries by themselves, we find that their aggregate GDP of $209.6 billion relates to an aggregate population of about 45.6 million, giving an average $4,597 per capita, as against $5,368 for Italy.

We will have more to say later in the chapter about the pattern of distribution of GDP in the oil-exporting countries. For the time being, all that concerns us in the present discussion is that average GDP varies widely between one country and another, and much more so within countries between the members of different socio-economic groups. Yet this overall average permits substantial resources for private and public consumption, investment, and arms imports in the relatively less-populated oil-exporting countries (Kuwait, Qatar, United Arab Emirates, Saudi Arabia, and Libya), and adequate resources for the larger populations of Iraq and Algeria. But it is impossible to determine the proportion of oil revenues that goes into each of the three categories of utilisation. This is not merely because of the unavailability of information on certain types of public expenditure, but also because a society treats the resources available to it as flows that come via different streams but converge on one reservoir, on which society draws for various purposes.

These reservations notwithstanding, we can say with complete certainty, based on obvious quantitative reality, that the oil-exporting countries could not have directed considerable resources to any of the three categories of expenditure, had they not taken over the power to correct oil prices in the early 1970s, and to raise their production as they did. The ability to shape policies and thus influence the volume of revenues earned is no doubt a blessing to the Arab countries, but some aspects of the utilisation of these revenues cannot be considered an unmixed blessing. As we will examine the most serious socio-economic implications of these revenues later on, we will only dwell here on the quantitative aspects of revenue utilisation: that is, on the magnitude of consumption (whether of domestic or imported goods and services), development outlays, and disposal of surpluses beyond these avenues of expenditure, leaving arms purchases out as they relate to security and therefore are not referred to explicitly in the national statistics. The quantitative aspects to be discussed will refer only to 1979. It would be misleading to put the whole decade under analysis, because during most of it revenues were substantially below what they were to become in 1979 and what they promise to be in the 1980s and probably the 1990s. Thus 1979 becomes a sort of prototype year, suggesting the minimum volume of revenues that are likely to accrue from oil exports for many years to come.[5]

Total consumption by the oil-exporting countries in 1979 — private, and public (recurrent) expenditures by governments (salaries and wages, rents, non durable supplies) — amounted to an aggregate of $90.24 billion, or about 43 per cent of their aggregate GDP, with private consumption accounting for about 27.4 per cent and public consumption for 15.6 per cent (see Table 4.3). The proportion of resources available to society going to consumption is not unduly high in relative terms by Third World standards, but once the absolute magnitude of GDP is taken into consideration, the outlay will prove quite large for countries at a low level of development like the ones under examination. To this phenomenon of serious implications must be added that of the very high level of imports, given the aggregate population of the seven countries under examination. This dual problem of high consumption and high imports needs some further discussion.[6]

The record of the 1970s reveals areas of over-use and misuse of financial resources, whether in the adoption of excessive

TABLE 4.3: Resources and Resource Uses of Seven Arab Oil-exporting Countries, 1979 ($million at current prices)

| | Iraq[a] | Kuwait[a] | United Arab Emirates | Qatar[a] | Saudi Arabia | Libya[b] | Algeria[a] | Total |
|---|---|---|---|---|---|---|---|---|
| Exports of goods and non-factor services | 21,448.5 | 19,480.8 | 10,099.3 | 3,588.4 | 54,569.3 | 15,926.3 | 10,098.2 | 135,210.8 |
| Imports of goods and non-factor services | 12,382.3 | 7,298.5 | 5,878.6 | 1,425.2 | 31,053.6 | 10,741.4 | 12,141.0 | 80,920.6 |
| Private consumption | 9,405.5 | 5,940.6 | 3,202.5 | 224.2 | 18,895.9 | 6,434.4 | 13,301.0 | 57,404.1 |
| Public consumption | 4,473.2 | 2,696.8 | 2,038.4 | 591.7 | 16,845.6 | 5,104.2 | 1,081.0 | 32,830.9 |
| Gross fixed capital formation | 11,176.8 | 2,501.6 | 5,110.7 | 797.5 | 23,212.2 | 6,041.5 | 13,798.0 | 62,638.3 |
| Change in stocks | | −22.4 | | | 1,445.0 | | 1,052.0 | 2,474.6 |
| Gross domestic product[c] | 34,121.7 | 23,298.9 | 14,572.3 | 3,776.6 | 83,914.4 | 22,765.0 | 27,189.2 | 209,638.1 |

Notes: a. Estimates by the Arab Fund for Economic and Social Development, AFESD. b. Source (a) contains data for Libya for 1978. However, the data shown in this table relate to 1979 and come from source (b) below. c. Source (a) contains an error of $10 billion under the heading 'GDP'; the error has been corrected here.
Sources: (a) The Secretariat of the League of Arab States, AFESD, and the Arab Monetary Fund, *The Consolidated Arab Economic Report, 1981*, Table 2/7, p. 259;
(b) AFESD, *National Accounts Country Tables* (Kuwait, 1980).

consumer-oriented styles of living, or in permissiveness in the design and costing of development works. The phenomenon is not one encountered only in oil-exporting countries, but in almost all the countries of the region. Private consumption is advancing fast, in no small part under the influence of TNCs and their sales promotion campaigns. These campaigns, conducted through the very effective modern mass media, create certain wants where none or few existed and transform these wants into pressing needs. They thus reverse the dictum which speaks of the 'sovereignty of the consumer' into one of the 'sovereignty of the giant producer'.

In addition to 'high consumption' (in the Rostovian terminology[7]), which threatens to arrive before the stage of 'high production' or 'high development' and which leads to high importation owing to the weak and limited productive capacity of the region, there is overspending on development projects; but more significantly we witness the building of projects of a very low degree of urgency. Cases of 'conspicuous investment' or showy development are abundant enough, in almost every country, to justify special condemnation. Here, again, the volume of oil earnings by the producing countries and the flow of funds into Arab non-oil countries must be blamed in part for the misallocation of resources, particularly in the presence of other programmes of much greater urgency and higher priority.

Apart from the very high level of consumption and the fact that it has been rising steeply through the 1970s, three very disturbing features are observable. The first is that per capita consumption varies sharply between oil-exporting countries and the other Arab countries, standing at about $1,979 and $557 for the two groups respectively. But it represents a very high proportion of GDP per capita for the second group, compared with the oil exporters: 89 per cent versus 43 per cent.[8] The fact that the proportion is low for the oil-exporting countries provides only illusory comfort, considering the very high level of GDP per capita in this group of countries. On the other hand, although the higher proportion for the non-oil exporters is to be expected, considering their low GDP per capita, its actual height is nevertheless very disturbing. This level is determined by the 'imitation effect' of consumerism, coupled with the availability of liquid resources. Some of these resources have been created by the earnings of Arab manpower working in the oil-exporting countries and making substantial remittances to their families back home. Another part represents flows from the governments of oil-exporting countries, part of which seeps through to the populations of the receiving countries.

The second disturbing feature is the heavy dependence on imports, in satisfying the high levels of consumption, which causes a serious leakage of purchasing power to foreign economies. Imports per capita are very high in the Arab region — being the highest in the world for a few oil exporters. Furthermore, more than 93 per cent of the imports come from non-Arab countries.[9] Hence the magnitude and seriousness of the leakage. (Exports, too, stand at a very high level — the highest per capita level in the Third World — obviously because of oil.) Together, imports and exports constituted about 94 per cent of GDP in 1979, against 60 per cent in 1970.[10] The combined effect of high consumption and high dependence on imports to satisfy much of the consumption is a legitimate cause for concern. The concern mounts even further when the total size of the external sector is considered, in contrast with the size of the domestic sector, and when other manifestations of dependence on the industrial economies are considered too.

In the present context, the combination of high consumption and high imports poses added difficulties for the integration of the oil sector with the producing countries' economies and with the region's economy considered as a unit. The combination leads to the intensification of the region's dependency on, and integration with, the industrial economies from which most of the imports come. The large volumes of consumption and investment could together be a powerful promotive force for the building of productive capacity inside the region. This potential is to a considerable degree wasted, however. (And this applies to the importation-versus-production of goods as well as of services.)

The third disturbing feature is the heavy dependence on oil revenues for the high levels of both consumption and imports. Aggregate consumption in the whole Arab region constituted 55 per cent of aggregate GDP in 1979. If oil revenues earned by all 12 Arab oil producers were excluded from GDP, consumption would exceed GDP by 12 per cent. The same calculation would reveal that consumption exceeds aggregate GDP by about 21 per cent if oil revenues are deducted from the GDP of the seven oil-exporting member countries of OPEC. Of course this is an absurdity, since consumption would not then be at all near its present high level. But the exercise puts the warning regarding consumption and consumerism into strong relief. This is because once consumption habits take deep root, it would be difficult to change them. The oil-exporting countries would then find themselves forced to produce more oil than it would be wise to do, merely to finance substantial (and probably expanding) consumers' goods and services.

Again, to extend the exercise, the imports must be considered in relation to oil revenues. Total imports for the whole Arab region for 1979, which stood at about $113.5 billion, represented about 39 per cent of GDP. If oil revenues of all 12 oil-producing countries were excluded from GDP, however, imports into the region would amount to over 40 per cent of the region's aggregate GDP.[11] Much more significantly, imports into the seven oil-exporting countries by themselves for 1979 amounted to $80.9 billion, while the non-oil exports of these countries for the same year amounted to $3.55 billion.[12] Without oil, these exports could at best finance only 4.4 per cent of the imports. It is as though a large proportion of oil revenue is generated in order to permit vast imports, thus further integrating the sector with the international, not the national economy. Furthermore, some 40 per cent of the imports are consumer goods and services, including an unusually high proportion of non-essential and luxury items where the oil-exporting countries are concerned. Obviously, the last statement involves a value judgment, and it would be more appropriate to note that some 60 per cent of the imports are goods and services for investment purposes[13] — tools, machines, equipment, and raw materials. The latter formulation would avoid controversy regarding the definition of what is a non-essential or luxury import.

However, the impressiveness of the magnitude of imports for development must be qualified in three ways: first, the fact that prices of imports are grossly inflated; secondly, that commissions on most construction and supply contracts are rarely less than 5 per cent, which leads to the exaggeration of the value of investment made; and thirdly that the insufficiency of local experience often permits serious and costly faults in design and/or construction to creep in, and it also permits investment in projects at a low level of priority in the overall strategy of development. (There is a fourth qualification which has been mentioned already — namely the leakage that imports constitute as they do not fall in the domestic cycle of economic life, but promote foreign economies instead.)

Though the allocation of oil revenues to the importation of consumer goods and services is substantial, and can be criticised to the extent that it encourages consumerism and weakens saving habits, the largest single allocation is to development, however vulnerable to criticism the course of development is.[14] This is made necessary by the state of underdevelopment from which the oil countries suffer and their strong desire to develop fast, as it is made easy by the fact that the oil revenues form a surplus in public hands — one which can be controlled and directed by central decisions by government. Were the oil industry in private hands, it would be impossible to mobilise as large a volume of resources from oil exports for use in development work.

As indicated earlier, no one part of oil revenues is earmarked for developmental investment, as against other parts for private and public consumption, for arms purchases, or for foreign aid. What matters instead is the allocations made, from the general pool of resources available, to the various uses to which society desires to direct resources. (These uses are consumption — private and public, capital formation, exports minus imports, and changes in stock.) The structure of resources use for each of the seven countries under survey is shown in Table 4.3. This table reveals that aggregate gross fixed capital formation amounted to $62.6 billion in 1979. A more complete picture of the volume of development investment would require the addition of current expenditures associated with fixed capital formation (not with the operation of the capital assets, but with their establishment), to the extent that national accounting restricts the record of fixed capital formation to the cost of capital assets by itself. But as we have no way of ascertaining the volume of the current outlays referred to, we will take the volume of gross fixed capital formation, namely $62.6 billion, to be equal to gross investment in 1979, and therefore to aggregate development outlay for that year, bearing in mind all the time that development outlays can also be made in the form of recurrent expenditures.

Aggregate gross investment for 1979 amounts to about 30 per cent of aggregate GDP. This proportion is very high by normal standards, but very small if seen in the narrow context of total resources available, current for 1979 and accumulated from earlier accruals, or in the context of the needs of the economies if accelerated development is to be sought earnestly. However, some such proportion is probably 'just right' for the countries in question, considering the many inherent constraints within society and economy which make the effective and non-wasteful utilisation of larger outlays impractical. We refer essentially to the shortage of skilled indigenous manpower; the difficulty of involving a larger Arab expatriate labour force than already is involved in the national economies (coupled with the domestic reluctance to bring about a larger involvement); and the inadvisability of bringing in a yet larger labour force from non-Arab Far Eastern countries like South Korea, the Philippines, India, Pakistan, or Sri Lanka.

There are serious socio-economic implications to the resort to a large expatriate labour force. But apart from shortages in indigenous skilled manpower and constraints on the resort to expatriate manpower, there are other factors which compress absorptive capacity. One of these is the necessarily slow pace at which more advanced, appropriate technology can be mastered, the generally inappropriate policies adopted for such mastery, and the danger of excessive resort to the importation of foreign technology both in its hardware and software forms. Another is the inadequacy of institutions — for financing, research, engineering for design, construction, quality control, and supervision, to name a few categories. Yet another is the costliness of development work. We refer here to the dual impact of inflation and the discriminatory pricing policy exercised by the suppliers of capital goods and technical skills in the advanced industrial countries, as a result of which development projects turn out to be excessively expensive. This means that the resources available stretch less far than they otherwise could.

The pattern and content of development emerging is one which puts much greater emphasis on the industrialisation of the oil sector, through the introduction and/or expansion of refining, gas treatment, or petrochemical manufacturing. This is defensible in all the oil-exporting countries, but particularly so in those with relatively small populations and with a narrow economic base, such as Kuwait, Qatar, the United Arab Emirates, and less so in Libya and Saudi Arabia. Iraq and Algeria present a clear case of a much wider base where economic diversification can be carried to a far extent, including the development of a robust agricultural sector. The needs and the possibilities of development are to a large extent reflected in the development plans formulated by the oil-exporting countries, where allocations to the various sectors indicate the social and economic preferences of the political decision-makers and the planners, and the pattern and content of development desired for the future.

The developmental allocations made in the 1970s as shown in Table 4.4 are considerably larger than those that were made for the immediately preceding generation of development plans. Late in the 1970s, with expectations of oil

TABLE 4.4: **Sectoral Distribution of Planned Investments, 1970-80 ($ Billion)**

| Country group | Agriculture | Manufacturing and mining | Building and construction | Electricity and water | Transport and communications | Health and education | Trade and financial services | Other service sectors | Total |
|---|---|---|---|---|---|---|---|---|---|
| Group I | 11.2 | 28.1 | 9.1 | 2.0 | 10.8 | 5.2 | 0.6 | 7.4 | 74.5 |
| Group II | 6.3 | 24.3 | 21.4 | 18.4 | 25.1 | 33.7 | 6.4 | 23.3 | 158.8 |
| Sub-total | 17.5 | 52.4 | 30.5 | 20.4 | 35.9 | 38.9 | 7.0 | 30.7 | 233.3 |
| Group III | 12.3 | 21.1 | 12.0 | 12.0 | 17.1 | 5.3 | 3.5 | 4.9 | 88.2 |
| Group IV | 3.8 | 2.7 | 2.2 | 0.8 | 4.2 | 1.2 | 0.7 | 1.1 | 16.7 |
| Total | 33.6 | 76.2 | 44.7 | 33.2 | 57.2 | 45.4 | 11.2 | 36.7 | 338.2 |

*Notes*: The composition of the Groups is as follows: Group I: Iraq and Algeria; Group II: Kuwait, UAE, Qatar, Saudi Arabia, and Libya; Group III: Bahrain, Oman, Jordan, Syria, Lebanon, Egypt, Tunisia, and Morocco; Group IV: North Yemen, South Yemen, Sudan, Mauritania, Somalia, and Djibouti.
*Source*: *Consolidated Arab Economic Report, 1981*, Table 2, p. 201.

revenues brightening further, many of the Arab countries (oil-exporting and others alike) began preparing more ambitious plans for the first half of the 1980s. The largest expansion was in the Saudi plan; the Saudi finance minister in March 1980 put total allocations for the third five-year plan (1980-5) at $267.9 billion, as against total allocations of $148.2 billion for the second plan (after upward adjustment).[15]

Though increases in the other plans were on a much smaller scale, together they constitute considerable additions to the investment allocations shown in Table 4.4. The realism of planning on such a huge scale in societies suffering fundamental shortages in technological, human, and institutional resources like the Arab countries must be seriously questioned, as the next section of this chapter will do. But one thing can be said in defence of the new plans for the years 1980-5, involving as they do an increase of about $150 billion over their level in the preceding decade, which is that they earmark a large and expanding proportion of oil revenues for development, thus placing a constraint on other non-developmental outlays.

The fact that the oil market in 1981 and the first quarter of 1982 (when this is being written) has been slack, and prices and revenues have dropped notably, will tend to confirm the tendency for greater discipline in spending and more moderation and realism in development planning.[16] It is necessary to add that the grim expectations of revenue by the oil-exporting countries will have their depressive effects on development allocations for the period 1980-5 of the non-oil countries, inasmuch as the drop in the level of revenues will no doubt influence Arab foreign aid commitments adversely. These commitments and disbursements have been a not-negligible part of the resource availability on which the non-oil countries have counted since 1974.

Another adverse effect of grim price and revenue expectations is the (possible or actual) scramble for markets, which in itself would also press the price structure and therefore revenues downwards. Under such conditions, and with rather rigid spending and importation patterns, the likelihood is great that the producers would want to expand production in order to maintain their desired volume of revenues. If the market is saturated, such a course of action would prove futile — indeed, it might prove counter-productive, as it would press prices downwards yet further. Whether it is depressed prices (given a certain volume of exports), or expanded production (given a certain price level) that finally absorbs the effects of the weakness of the market, the final result is reduced revenues at least in the short run. In the conditions that have prevailed between the price increases of 1979-80, and the considerable weakening of demand until the spring of 1982, it has been prices and production that together have taken the shock. This is because of the building of immense stocks in 1979-80 (estimated at over 5 billion barrels), which has permitted the industrial countries to put pressure on prices *and* to reduce their purchases by drawing on their stocks. Rather than expand production in order to maintain revenues at a lower level of prices, the producers have had to reduce production in order to stop the erosion of prices.

Before we move on to discuss the last issue with which the present section is concerned, it would be appropriate to indicate the five sources of pressure on OPEC, therefore mostly on Arab oil. The first, as suggested above, is the huge stocks which are being depleted in order to avoid the holding of excessive highly-priced stocks. The second is the speeding up of substitution (mainly of coal) for oil, under the pressure of the rise in oil prices in 1979-80. The third is the intensified effort at energy conservation in the industrial world. The fourth is the expansion in the production of non-OPEC oil, a larger volume of which now goes into international trade. The fifth and last source of pressure to mention is political: the deliberate attempt to weaken OPEC if not bring about its total disintegration. The instrument opted for by the governments and oil companies of the Western countries is to let OPEC oil get the full impact of all four factors mentioned above. The weakening or disintegration of OPEC is not a secret objective; nor is it a menace which OPEC members

*Oil as Engine of Development*

imagine. It is an objective sought by some Western quarters since 1973.[17]

This section of the chapter deals with the financial implications of oil policies. The last implication to be discussed is the accrual of what has come to be called the 'surpluses' of the oil-exporting countries. These are nothing but the unspent financial resources that these countries have received from the sale of their precious, non-renewable asset which is oil. But we need not dwell on this point any longer, as it has received many references in writings on oil matters. There is even some questioning of the propriety and advisability of including oil revenues among the components of gross domestic product, since they arise from an exchange of one asset for another, rather than from a continuous flow of income deriving from the production of a commodity that is renewable. However, there is one aspect of surplus-creation that should be referred to, even if briefly. This is the complaint by the large consumers-importers of oil that the upward-adjusted prices of oil have resulted in a 'massive transfer' of real resources from the consumers to the producers of oil. This complaint merits some discussion.

The first point of explanation is that there is, in return, an obvious transfer of real resources in the opposite direction: from the producers to the consumers. The financial resources only flow to the producing countries because of the flow out of these countries and into the consuming countries of a most strategic resource: oil. This resource is vital to the economic performance and growth of the importing countries, as well as to their military capability and security. The critical function of oil cannot be questioned convincingly either by the economists or by the politicians/strategists of the importing countries.

In the second place, the financial resources flowing into the oil-producing countries are large in volume only in relation to the total resources available to these countries, not in relation to those available to the advanced industrial countries which voice most of the complaint. In addition, the oil revenues are critical for the development of the oil-exporting countries under discussion, all of which are poor in other resources and have seriously under-developed economies. On the other hand, the industrial countries have advanced and diversified economies, a high level of economic and technological performance, and the structures and institutions that go with a high stage of development.

Thirdly, the resources earned by the oil-exporting countries in the most part find their way back to the advanced countries. The 'return journey' is made either in payment for goods and services imported for investment and consumption purposes, or by investment or placement in the economies and financial institutions of the industrial countries. (Even when the oil-exporting countries use some of their financial resources to import goods and services from the socialist or developing countries, most of the resources thus directed end up in the Western industrial countries, to pay for imports into the socialist or developing countries themselves.[18]) It is relevant to add that the financial resources returning to the Western industrial countries have had to be paid for goods and services that have been becoming costlier continually since the beginning of the 1970s. The burden of inflation that the oil-exporting countries have had to bear is much heavier than that which the oil-importing Western countries have had to bear as a result of rising oil prices, if only because the real price of oil has not risen commensurately with the real prices of goods and services imported by the oil exporters (see Table 4.1). In addition, the comparison between the ability of the two groups of countries to absorb inflation is vastly in favour of the advanced industrial countries.

Finally, that part of the surpluses directed to the financial and money markets of the Western countries for placement in banks (and it is by far the largest single part[19]) has suffered a substantial erosion in its purchasing power.[20] And to the extent that such placement is made in US dollars or Eurodollars, it has additionally suffered from fluctuations in the value of the dollar against other major currencies (again as Table 4.1 showed). At this point the basic question should be, not what burden the industrial countries are bearing because of rising oil prices, but why the oil-exporting countries should in the first place produce and export more oil than they need for their national and regional consumption and development. As indicated earlier in Chapter 3 (during the discussion of production policies), most of the Arab oil exporters can manage with a substantially lower level of production, were it not for their acceptance of a large measure of international responsibility. It is a painful irony that the oil exporters are blamed rather than thanked for the shouldering of such responsibility.

The discussion of the issues associated with the surpluses of the Arab oil-exporting countries can be undertaken without detailed consideration of the size of these surpluses. This is what we have done so far. But perhaps some reference to the estimates that have been made of the surpluses will be in order for the rounding of the discussion. Many such estimates have been made, mostly by leading banks in certain Western countries. As a result, 'Confusion reigns over the size of OPEC surplus', as an authoritative oil periodical has put it. In a comparison of the estimates made by seven leading institutions, including the Organisation for Economic Cooperation and Development, OECD, the estimates vary widely.[21] For instance, when the estimates were made in the early summer of 1981, the forecasts for OPEC members' current account surplus (after official transfers) for 1981, varied from a low of $67.6 billion (First National Bank of Chicago), to a high of $109 billion (OECD). The range of forecasts for 1982 was narrower, stretching from $49.2 billion (First National Bank of Chicago) to $69 billion (Morgan Guaranty Trust). Consequently, the estimates of cumulative OPEC net foreign assets also vary widely. The high and the low estimates of cumulative OPEC net foreign assets for 1979, 1980, and 1981 stand as follows (in $ billion): 1979: high 224/low 196; 1980: high 327 (forecast March 1980)/low 279; 1981: high 407 (forecast March 1980)/low 346.

The assets of five Arab oil-exporting countries which have accumulated substantial reserves were estimated at $178 billion by the end of 1979, and were forecast at the time to reach

$275 billion by the end of 1980.²² (The two countries excluded are Algeria and Iraq. The first, in fact, had a much larger foreign debt at the time than its foreign assets, and the latter did not build up substantial reserves because of its own considerable needs for foreign exchange and its relatively large population and broad-based economy with large absorptive capacity.) The size of these foreign assets is a source of fear to the owners – not merely because of the continuous loss of value through inflation and the erosion of the purchasing power of the assets, but also because of the danger that the assets, or whatever part of them is in the United States, could be frozen if serious political differences arose between the owners and that country.

On balance, therefore, it can be said that the financial implications of Arab oil policies are not an unmixed blessing. The positive aspects of the inflow of relatively large oil revenues are obvious; they include the financing of pent-up demand, especially of the lower-income strata of society; the financing of expanded development outlays; the financing of increased imports; the financing of arms purchases for defence. However, each of these positive aspects has serious drawbacks and calls for significant qualifications. As these qualifications relate both to the present section and to the one to follow, the identification of the major qualifications is made in both sections.

## The Developmental Impact of the Policy of Control

Some reference has been made to one major use of oil revenues, which is to finance development, primarily in the oil-exporting countries themselves, but also – though to a much smaller extent – in the non-oil Arab countries. Thus, the developmental implications of the policy of control and by extension of oil revenues, are both national and regional. Both categories of the implications will be discussed in the present section, though the allocations of resources to foreign aid by OPEC members, among whom the Arab countries constitute by far the major donors, will be discussed in the next, and last, chapter of the book, where oil will be examined in its international context. In other words, the developmental implications of Arab oil, which seem to fall in the area of international relations to the eyes of the non-Arab, are considered a regional matter to the Arab believers in Arab Nationalism (to whom the whole Arab region is the habitat of the Arab Nation, and the Arab peoples in the 21 states are but one Nation). Our approach will involve a compromise: the inclusion of regional development in the present section, where 'national' (that is country-by-country) development will be discussed; and the inclusion of financial flows to non-oil Arab countries in the discussion of foreign aid, which will occupy part of the next chapter, in order to obtain an integrated picture of Arab contributions to Third World countries as a whole.

Table 4.4, which presented a picture of planned allocations of funds for development purposes, detailed by Arab country groups and by sector, appeared in the last section. The other forms of utilisation were consumption (both private and public, including defence outlays), and investment and financial investment or placement of funds in banks. Now we turn to a closer examination of the content and meaningfulness of Arab development — both national and regional — and of the true sense in which the oil sector has served as an engine of development. We will also assess the extent to which the oil sector has been effectively integrated with the national and regional economies. The discussion will attempt to show both the positive achievements and the shortcomings of development and integration, and thus suggest avenues for correction. However, though of great significance, this discussion will be brief, since the present writer has undertaken fuller examination of Arab economic development and of the integration of the oil sector with the Arab economies elsewhere, and does not feel justified in repeating much of the same material here.[23]

The role of the oil sector as an engine of development has changed in breadth and emphasis over the years, to the extent to which the sector has been closely and effectively related to development effort, and to the extent that it has been closely and effectively integrated first with the economies of the oil-exporting countries, and then with the economies of the other Arab countries. The two criteria are not separate, inasmuch as the contribution of the oil sector to development effort comes about via one of the aspects of integration, as we shall see presently.

Essentially, assessment of the extent and quality of integration must be sought primarily with respect to the locus of the decision-making power in matters relating to oil policies. (See Chapter 2 above where this aspect was discussed under the heading 'The Policy of Control'.) It is this shift in the locus, from foreign concessionary companies to national governments, that has made possible the exercise of power relating to all other oil policies and decisions. In the second place, integration can be reflected in the emphasis on the role of the oil sector as a contributor to the reservoir of financial resources at the disposal of the economy. It is in this capacity that the oil sector made its greatest contribution to development in the decades preceding the 1970s, particularly between the early 1950s (when the equal sharing of net export profits between concessionary companies and national governments was initiated), and the autumn of 1973, when the policy of control by governments was consummated. And finally, integration can be assessed in the extent to which the oil sector has succeeded in promoting development by bringing about significant structural changes in the economy, not just by providing development finance.

This last aspect of the role of oil, namely that of prime mover or engine of development, permits oil to make its maximum and most meaningful contribution to development. The oil sector thus witnesses a change in emphasis from an originator of finance, to a dynamic, leading sector which permits the creation of vital linkages, first between itself and other sectors and activities and capabilities, and then at a further remove between the first generation of sectors and activities influenced by the integration, and second and third generations.

The linkages can be made backwards, that is with capabilities or activities lying behind upstream operations and supporting them (like industries and skills needed in prospecting and exploring for oil as well as in oil production and transport); or they can be made forwards, that is through the development of activities and capabilities falling within and required by the wide sphere of downstream operations (like refining, petrochemical industries, tankerage, and distribution). At further removes, these forward linkages include new or extended activities in sectors and industries that become users of inputs provided by the refining and petrochemical industries, and in supportive, ancillary, and generally-associated industries and activities.

One major beneficiary of all this interconnection and ramification of activity is research, training, and the acquisition of technological capability; and this beneficiary, in turn, becomes the benefactor to the economy as a whole, even in areas of activity distant from the oil sector, through an obvious process of seeding and diffusion. The forward-stretching chain can indeed be very long and involve an enormous volume of investment and employment, considerable diversification in the economy (particularly within manufacturing industry), an impressive widening of the range and quality of skills in society, and finally substantial contributions to gross domestic product.

The Arab oil sector has achieved the first two phases of integration, with what they involve by way of control of oil policies, particularly in the broad area of upstream operations, and by way of the generation of substantially larger volumes of revenue which it has been possible to divert to national and regional development (and to certain other needs associated with consumption and security). Likewise, the financial resources that have become available thanks to the policy of control, have made possible the provision by the Arab countries of significant foreign aid to Third World countries and to international organisations for development purposes. Indeed, the record of the Arab oil-exporting countries in this respect has been distinctly creditable; and to this extent they have made what must be considered good use of the oil revenues that have accrued to them since the autumn of 1973, as far as the proportion of revenues directed to development is concerned — national, regional and international development alike.

At the national level, the developmental allocations made possible by oil revenues are capable in themselves of producing a respectable rate of growth, and of promoting development in the wider, deeper, and more meaningful sense of the term, if the necessary conditions are satisfied. The enquiry into the limits of such a true developmental performance will be undertaken shortly. For the time being, we will dwell on the question of growth made possible by oil revenues. The magnitude of the rate of growth is subject to a number of factors that make it difficult to estimate growth on the basis of the investments made. One of these is that investment in social and economic infrastructure (education, health, roads, water systems, and the like), which receives the largest proportion of resources, does not get translated readily into growth, owing either to the long gestation period necessary before the investment bears measurable economic results in terms of additional national product, or to the fact that its influence is diffused over the whole economy and society, and therefore is not directly attributable to the sector in which the investment is initially made.

Another factor is the disproportionately large influence that oil prices and the volume of oil production have on the rate of growth, whether upwards or downwards. As a result, growth is subjected to unpredictable occurrences lying often outside the power of the national economy. Furthermore, prices are often changed by government decision, which also makes the rate of growth move in isolation from the relationships between investment and output (what is known as the marginal or incremental capital/output ratio). Finally, there is the factor that the oil sector, in its capacity as the largest single contributor to GDP, can add considerably to GDP without any additional fixed investment (that is, with a very small capital/output ratio or capital coefficient), so long as the extra production falls within the established production capacity of the country concerned.

Be all that as it may, the aggregate investment made in 1979 by the oil-exporting countries was capable, on its own, of producing a rate of growth in output of some 10 per cent, apart from the physical possibility of a different rate associated with changes in production and/or export prices for oil. (As it turned out, 1980 marked a real rate of growth in GDP of about 14 per cent over 1979, thanks mainly to a rise in the real price of oil of about 14 per cent also, while the volume of production fell by 9 per cent.[24] (Growth in non-oil sectors, and possibly in productivity, must not be overlooked as a compensatory factor.) Having given some attention to the rates of growth achieved or possible, we should hasten to add that not much should be read into such rates, or, for that matter, into growth performance by itself, particularly in oil-exporting countries, since growth can be fast, while true comprehensive development is slow-moving. (Conversely, a drop in the rate of growth does not mean that development is regressing.)

So far we have referred to two criteria or manifestations of the integration of the oil sector with the Arab economies. The third criterion is the extent to which oil serves as an engine of development, through promoting backward and forward linkages, and leading to the several constructive aspects of economic and social transformation to which we drew attention earlier.

A careful examination of the course of development in the 1970s, especially since 1973, reveals mixed results. At first glance, it can be seen that an ambitious course of development was charted by the oil-exporting countries. The Arab non-oil countries also accelerated their development efforts, thanks to the contributions made by the oil countries (to the tune of $4 to $5 billion annually on the average) in loans and grants, and also thanks to the remittances (estimated at $3 or $4 billion a year) sent home by the three million Arabs working in the oil-exporting countries.[25] These financial flows combined are the equivalent of about one-third of the volume of gross fixed capital formation of the countries that fall into Group III and IV (see Table 4.4 for a grouping of the Arab countries and a listing of the members of each group.) However, this comparison is not very significant, since not all the official transfers or the private remittances are directed to investment back home. Defence outlays constitute a large part of the transfers, and consumption outlays of the remittances. On the other hand, the fact that much of the transfers and remittances goes to a small number of countries (mainly Egypt, Jordan, Syria, Lebanon, and North and South Yemen), means that they have a positive effect on the receiving countries, by enlarging the flow of investment financing and generally activating the economies.

Again, a first glance at the impact of oil revenues on development reveals that this impact has been felt at the regional level (or at the level of the Arab economy viewed as one entity), apart from the national level. The outward manifestation of this regional impact is the development of many areas of interlocking relationships. These include financial and manpower flows across national frontiers, to which reference has just been made. In addition, they include the formation of many regional institutions, engaged in joint production, advisory work, co-ordination, but above all joint ownership, within the public and private sectors, under the umbrella of the League of Arab States and the Council for Arab Economic Unity, but also outside this umbrella.[26] Together, the joint projects and programmes that are engaged in direct activity (in the sectors of industry, agriculture, finance, mining, and metallurgy — to name the leading cases), plus the development funds whose scope of activity covers the whole Arab region, have an aggregate capital that exceeds $40 billion at the time of writing. This large and ramifying body of regional institutions and activities is currently called the Joint Arab Economic Sector, JAES, and it has registered most of its expansion during the 1970s, and especially since late 1973, thanks mainly to the marked expansion of oil revenues flowing into several Arab treasuries.

The positive contribution of oil to development can thus

be seen to have flowed in three streams. These are: national development inside the oil-exporting countries:[27] national development in other Arab countries promoted directly through financial flows made possible by oil revenues, or through manpower and remittance movements, in turn made possible through the economic activity generated by the development work based on oil revenues; and regional development (that is, development of the Arab economy as one integrated entity) through the establishment and expansion of joint projects, programmes, and activities within the broad framework of the JAES.[28]

The convergence of these three streams of development in the Arab region justifies the conclusion that the oil sector has been instrumental in promoting development, and that the results of such promotion are both attributable to oil, and tangible. This must be borne in mind if the critique of Arab development, and the comments that will be presently made on the course and content of development achieved, are to be understood in true context. If the critique sounds harsh, this is because of the feeling that the efforts and resources directed to development should have produced more creditable results.

It is necessary at this point to point out that the results are quite creditable if measured in terms of rates of growth achieved, or indeed of intensification and diversification of investment.[29] The intensification is attested to by the vast absolute size of development plans in the Arab region as a whole, but particularly in the oil-exporting countries, by the steep rise in the volume of investment allocations made after 1973, and; indeed, by the tangible rise in the rate of execution of projects. Even after allowing for the effect of inflation on the magnitudes involved, the statements made remain valid.

The diversification is less in evidence because of the overwhelming importance of oil in the sectoral structure of the Arab economy as a whole. However, there is greater diversification and sophistication within sectors, in the sense that more and newer manufacturing industries are being introduced, with greater variety of products and complexity of techniques. Likewise the sectors of transport and communication, construction, banking and finance, trade, hotel-keeping — to name the leading cases — have all witnessed quantitative expansion and qualitative change in their technologies and diversification. Agriculture is the distinctly poor relation in this general picture, with the lowest rate of growth achieved in the past two decades among all the sectors, in spite of the obvious economic and social significance of the rural sector broadly considered.

But the record in terms of investment, diversification within the sectors, and growth, is not the area where more creditable results should have been expected, as indicated a few paragraphs back. That area relates to meaningful development, as against mere growth. It is maintained here that such development has been much slower in unfolding than growth has been in materialising. This is but one of a number of major paradoxes or contradictions that can be discerned in the process and record of development in the Arab region, but particularly in the oil-producing and exporting countries. However, before we identify these paradoxes and discuss them, we should return to the role of oil in promoting development through backward and forward linkages.

The extent to which this role has been discharged in the oil countries has already been noted and evaluated in Chapter 3, where upsteam and downstream operations were examined. It was noted then, in the context of upstream operations, that exploration activities were only partly integrated in the body of skills and capabilities of the oil countries, and were therefore only marginally developed under the impetus of the take-over of control. In other words, the backward linkage in this area was still weak, in the sense that the capital goods and the skills required for prospecting and exploration were still largely imported: the linkage was therefore mainly with the external economies. On the other hand, the production capability (like the transport and marketing of oil) was acquired to a much greater extent, judging by the proportion of these operations undertaken by the nationals and institutions of the countries concerned. (To the extent that other Arab nationals are also employed in the various aspects of these operations, the linkage has been established with the regional economy, and is not a leakage as in the case of exploration.)

Forward linkages in the form of refining and petrochemical industries are being forged at a quick pace and through substantial investment, as we have had occasion to show in Chapter 3 and in data on planned investment (Tables 3.8 and 4.4). But here again, oil is just beginning to discharge its function as an engine of development. This is because the vast expansion in refining, and the initiation of petrochemical industries, are in their first phase, being post-1973 developments. Furthermore, the capital goods and the skills involved in the industrialisation of the oil sector, are virtually all imported from the advanced industrial countries. Finally, what is being achieved is only the establishment of the first generation of forward linkages, with subsequent generations being still matters of expectation and hope. It will be some time yet before the oil-exporting countries (and the region as a whole) will have developed the capability to absorb and utilise much of the products of the refining and petrochemical industries as inputs in new industries and activities.

To make these comments on the shortcomings or limitations of backward and forward linkages is not to belittle what has so far been done. Indeed, the performance of the Arab oil-exporting countries is worthwhile in this respect, considering the short period of time which these countries have had to develop the linkages, with all that this involves in the development of national capabilities. The diversity and complexity of the capital goods and technical skills required make it necessary for a long time to elapse before a reasonably satisfactory acquisition of such capabilities is achieved.

Broadly speaking, the industrialisation sought by the oil-exporting countries is a sound objective, particularly for the small-population, narrow-based economies. But a number of qualifications have to be inserted within this broad approbation with respect to the strategies and policies adopted and the avenues of action pursued. The criticism of national strategies and policies is essentially that these are not being formulated in such a way as to promise to lead soon to healthy, regionally self-reliant development. This weakness which is inherent in the conceptionalisation, the content, and the course of Arab development will become clearer to see in the discussion of the paradoxes to which reference has been made. Five of these will occupy us here.

The *first* is the possibility for an economy of achieving a high rate of real growth in national product and reaching a high income or national product per capita, while the level of development remains low. This paradox can be encountered in many Third World countries, but can be seen in its extreme form in the oil-exporting countries. The explanation in this latter case lies in the fact that oil exports make the major contribution, and a substantial one, to GDP, thus permitting a high per capita GDP. Indeed, the oil-exporting countries include one or two with the highest per capita GDP in the world. Yet the economic and technological performance of these countries does not entitle them to belong to the advanced economies. Their national product and financial resources notwithstanding, they remain part of the developing group of economies.

There is nothing to condemn in this contradiction by itsself. But it calls for concern to the extent that it is often accompanied by a false sense of satisfaction with the performance of the economy measured in terms of GDP per capita, satisfaction based on a confusion between growth and development at policy-making level. Furthermore, frustration might emerge at the popular level of awareness, to the extent that the statistical indicators of high income averages and substantial financial resources are not adequately mirrored in expanded areas of cultivable land, more and better schools, decent living conditions, more abundant opportunities for rewarding employment, or a more egalitarian pattern of income distribution.

In fairness, it ought to be emphasised that the oil-exporting countries are striving to provide their citizens with all those concrete essentials that money can buy, like education, health services and housing. But money cannot buy development, or endow the high GDP per capita with the meaningfulness of development for a society. Development must be preceded and accompanied by appropriately-oriented fundamental transformations in economic, technological, social, and political endowments, capabilities, and institutions. Only thus will society become the milieu that is capable of providing the economy with a great body of the ideas, knowledge, skills, attitudes, and institutions and forms of organisation that are essential to the operation of this economy with efficiency and rising productivity. In brief, it is the men and women of a country, duly skilled, motivated, and organised, acting within a framework that provides the appropriate mix of incentives and constraints, that can bring about development.[30]

If this is correct, then the situation of all the oil-exporting countries is still far from one which is conducive to tangible and sound development in as short a period of time as most of the leaderships and the populations desire and hope. This is basically because some of the necessary conditions are not being satisfied, particularly that of intensive involvement by the nationals and their readiness — even eagerness — to learn and perform many of the skills that a modern economy requires. Heavy dependence on expatriate labour, which began as an unavoidable necessity, has become an addiction. The ease with which nationals can make substantial fortunes (or at least the 'supposed or perceived ease') has resulted for many in a separation between effort and reward, and has weakened and damaged their work ethic.

Furthermore, the abundance of financial resources (including foreign exchange), has led to an excessive and dangerous permissiveness in contracting for projects and in the importation of goods and services of all types and for all purposes. Again, while this is unavoidable at an early stage of economic development, policies and measures must be formulated and taken in order to reduce gradually the leakage outward of effective demand by building internal productive capacity to substitute increasingly for imports. And, to the extent that this is not possible in the oil-exporting countries themselves, the promotive effect of the new purchasing power must be substantially directed towards other Arab countries where the building up of improved and/or expanded productive capacity is more readily attainable. It is quite possible that such developments will in due course take place. Yet development strategies and policies as they can be clearly seen at the present, do not seem to lead to active correction of the existing pattern, but to its continuation.

The *second* paradox or contradiction to observe is the attainment of high levels of national product per capita and high rates of growth, simultaneously with some serious economic and social imbalances.[31] This is as true of the oil-exporting as of the non-oil Arab countries. Indeed, it is true of many Third World countries, though in different degrees. Four areas of imbalance deserve special mention here. The first is the slowness of the process of sectoral diversification. This is observable particularly in the oil countries, where non-oil sectors have shrunk in relative importance; but it is also observable in the region as a whole, with industry still making a very modest contribution to domestic product (some 7 per cent as a general average), and with agriculture registering very little or even negative growth.

On the other hand, certain service sectors that are not particularly promotive of development (involving speculation in real estate and shares and stocks, middlemen's services, entertainments) are increasing their share in domestic product. The growth of the government sector (and of its contribution) cannot by any means be wholly considered an unmixed blessing, considering the padding of the civil service and the disguised unemployment that that creates, and considering the nature of some of the 'services' which governments offer that the population would be happier to be deprived of. The Arab countries have their due share of these universal developments. But considering the shortage of skilled manpower and/or resources, they can ill afford the misallocation of

*Oil as Engine of Development*

resources involved.

A second area of imbalance is that between the countryside and the urban centres, or between the capital and other cities, and between socio-economic groups. The imbalance can be seen in the different levels of well-being and conditions of life. At its starkest, it is encountered between the poorer parts of the countryside and the poverty belts (the *bidonvilles*) of cities, and the rows of villas in suburbia. It cannot be seen in bank accounts because of the discretion of bank managements, but it can be perceived. To a certain modest extent, this form of imbalance is disguised, thanks to the social services which are being offered by governments, and to the opportunities for education and social mobility that are available. But a massive part of the imbalance still exists and cannot be disguised.

The third aspect of the imbalance is the widening in the wealth and income gap in virtually all the Arab countries, as development proceeds, but particularly in a number of the oil-exporting countries where the possibility for amassing huge fortunes is much greater than elsewhere.[32] The Arab region is not unique with respect to gross inequality. But the widening gap is probably more glaringly evident than elsewhere. The fact that serious and effective corrective measures are not being taken in most countries (if they are at all contemplated) makes the phenomenon more exposed to observation and criticism. Its adverse implications for economic motivation and for stability and social cohesiveness, must be seriously considered. We must add in the present context that the oil-exporting countries, with one or two exceptions, have lightened the tax burden on their citizens considerably. Apart from the fact that this means total tax-relief for the well-to-do and the very rich, it also spells problems for the future when the governments realise that taxation is necessary.[33]

Furthermore, the gap is widening among countries, as it is within countries. In this respect, there is considerable scope for correction through a better allocation of resources between the capital-abundant and the capital-short countries. Rewarding economic opportunities are not in short supply, although certain institutional prerequisites have to be adjusted for them to become more inviting.

The fourth and final area of imbalance to mention is the persistence of substantial unemployment, in spite of a satisfactory record of economic growth. Again, this phenomenon is near-universal in Third World countries. But it acquires special seriousness in the Arab region, with massive development work and brisk economic activity going on in most countries. This is a clear case of misallocation of resources, arising from shortages in certain skill areas and abundances in others (and, naturally, abundance in the ranks of the unskilled); insufficient vocational and technical training and/or insufficient attraction to such training in several countries; inadequacy of arrangements to improve inter-country flows (although private initiative has resulted in substantial manpower flows across national frontiers, as already indicated); the wrong choice of technologies or areas of activity; and above all the insufficiency of effective concern with the problem of unemployment in those decision-making centres that deal with development and planning questions.

To go back to the identification of paradoxes, we encounter the *third* of these. This is the actual shrinkage of popular participation in the discussion and elaboration of political and economic decisions, and in the shouldering of responsibility for such decisions. This has been happening increasingly although the drive for sound and correctly-oriented development requires wide and intensive popular participation, even apart from the political and social grounds on which the case for political freedom and democracy stands very firmly. To restrict the case to economic considerations solely, we see that the transformations which are essential for the unfolding of comprehensive development are so multi-faceted and so pervasive, and they involve virtually everybody to such an extent, that participation becomes a very fundamental and necessary condition.

It hardly needs to be stressed that energetic development calls for and requires the efforts of all those who fall within the confines of the labour force, to whatever level of responsibility and skill they belong. It therefore requires their firm commitment to its objectives and policies, and to the tasks it calls for. However, where there is no participation, there can be no firm and honest commitment. Furthermore, unless there is wide and acceptable participation in the rewards and fruits of development — that is, in the distribution of the national product it puts in the economy — the commitment will be increasingly eroded and undermined.

Most of the Arab countries have fallen into the misconception that development is a process which is carried through by the political leadership and its specialised technical agencies, from the stage of conceptualisation and the formulation of strategies, plans, and policies, on to execution and follow-up. The people at large are seen as the instrument of execution. Indeed, serious follow-up and evaluation have been undertaken publicly only rarely in the past quarter-century during which planning has spread to the whole of the Arab region. The people, who are meant to be the beneficiaries of the development outlays and efforts, and who bear the burden of development work, are rarely, if ever, asked to express their opinion of the degree to which these outlays and efforts have succeeded in achieving the objectives initially set for them. Those who have been able to see the limitations of the achievements, and the magnitude of the resources and time lost in the pursuit of false or superficial development, have on the whole not been able to make their voices heard above the din of applause for the achievements. Needless to say, much of the applause in most of the countries has been officially solicited.

The *fourth* paradox to stress is the intensified dependence of the Arab economies in general, and the oil-exporting economies in particular, on the industrial world, while it is widely assumed and declared that development is making progress, and that the Arab region is acquiring increasing economic strength in the world economy, thanks to the economic and political muscle that oil has given to the Arab region. The intensified dependence takes many forms and can be seen in many areas of the life of Arab society and economy. It is to be seen in the patterns and styles of life and thinking, in technology, in the acquisition and diffusion of information, as in political and international relations. If we focus on economic life alone, we see the dependence in the rising ratio of imports plus exports to GDP. This ratio was 60 per cent in 1970, but has risen to 92 per cent in 1980 (94 per cent in 1979).[34] The imports include a wide assortment of goods and services, both for investment and consumption. The latter include substantial food imports (wheat features on the import list of every single Arab country), and an inordinately high proportion of luxury or less-essential goods and services. Over three-quarters of the imports come from OECD countries.

On the other hand, the exports include a much narrower range of goods and services. Oil (crude, with a modest share for refined products) constitutes 90 per cent of total Arab exports for 1980, but 98 per cent of those from the oil-exporting countries. Again, OECD countries receive over three-quarters of the exports. The implications of the basic features of imports and exports have very serious significance for the health as well as the security of the Arab economies. These features include not only the volume of the imports and exports and therefore the size of the external sector, but also the composition of the imports and exports, and their origin and destination respectively.

It might be argued — indeed, it frequently is argued by some Arab analysts but more so by some leading officials — that what is here described as serious and far-reaching dependence on the industrial world is in fact a case *par excellence* of *inter*-dependence. This is true, but only superficially. Account must be taken of the relative power — economic and political, and, by projection, strategic and military — of the two parties to the 'inter-dependence'. (Indeed, given the *im*balance of power, the term inter-dependence is in essence a euphemism.) Of strong and direct relevance, as far as imports are concerned, is the exposure of the Arab region to the power and attitudes of the advanced industrial world with respect to the control of food, technology, and armaments. The size and concentration of Arab surpluses in a few Western countries is another aspect of the serious exposure. What makes the dependency and the exposure all the more disturbing is that they arise in large part from growing and pervasive consumerism coupled with ill-conceived and unsoundly-oriented development.

One powerful reason why such development produces and magnifies the dependence is the failure of the Arab economies to direct their resources and energies towards increased regional self-reliance. Were this failure to be assessed properly and to be effectively corrected, many of the goods and services imported could be produced in the Arab region, through a sounder use of the region's major export: oil. This resource could serve the purpose of promoting more balanced and diversified economies, and these would be capable of restructuring their imports and their exports in a fashion that would reduce the intensity of the dependence. It is only when the different aspects of exposure are tangibly corrected that a proper and acceptable state of inter-dependence can be approached.

The *fifth* and final paradox to consider here is the excessively narrow focus of Arab development, both in the oil-exporting and non-oil countries.[35] We refer here to the restricted country focus adopted, rather than the much broader regional focus, in spite of the fact that theoretical analysis and empirical evidence both show that the drive for development benefits greatly when it has a regional dimension. The justification for the concern with such a dimension does not merely lie in the many non-economic ties that bind the Arab countries together — shared history, shared culture, shared dangers and hopes — but also in hard-headed calculation of economic interests and potential.

In brief, the Arab region constitutes the strategic economic depth for each of its components, and much can be gained by development within each country if it took into account certain regional requirements and endowments. Thus, the strategic economic depth provided by the region can be the expanded market, the scope for horizontal or vertical integration between productive facilities, the availability of larger manpower and material resources and the possibility of a more rational allocation of these resources among the region's countries, the provision of a solid economic base for national and regional security, and the creation of a strong protective capability for developmental achievements.

Thus, serious concern for the development of each of the Arab countries involves, minimally, the elimination of contradiction between the objectives, pattern, and courses of the various development processes — a contradiction which is becoming increasingly threatening with the tightening and narrowing of country outlooks. As a second stage, correction of the present state of affairs requires the deliberate establishment of harmonisation and internal consistency among the various processes and courses of development in the various countries, that is, movement to a situation where it becomes possible to weave together certain aspects of country development plans or programmes and projects, thus achieving interaction and complementarity. This complementarity should and can go beyond mere joint financing of projects, haphazardly emerging and scattered, to involve joint production and joint labour participation, for the benefit of a common market.

Finally, the concept of Arab economic development can, and should rise to the stage whereby certain sectors, programmes and projects which form the scope for joint Arab economic action, are systematically included within a regional plan for the Joint Arab Economic Sector, JAES. This would be the highest achievement, within the present framework characterised by the predominance of national sovereignties.

It is this last stage that was envisaged in the Strategy for Joint Arab Economic Action which had been formulated in 1979/80 and was approved by the Arab Heads of State at the 11th Summit Meeting held in Amman, Jordan, in November 1980.[36] The Strategy document stipulated, among other things, for the mechanisms necessary to expand and strengthen the JAES. The most notable of these mechanisms was a plan for the JAES, and the finance required for the joint programmes included in the plan. The rationale for the strategy, and for the acceptance of its tenets by the heads of state, was the conviction that the individual processes of development in the various countries would benefit from the acceptance and adoption of certain regional developmental tasks. While the net benefit to the individual countries is demonstrable, it is equally demonstrable that there are certain regional objectives (in the areas of development and security) that would also be well served through such a regional outlook, with the joint tasks and obligations it involves. Finally, self-reliant development was conceived to be possible only within a regional framework.

The discussion of development at the country and regional level in the present section of the chapter is closely tied to oil policies, and the tie goes well beyond the revenues accruing from the export of oil which can be utilised in the financing of development. We have tried to indicate the frontiers of the wider context of the connection between oil and national and regional development. In the course of the discussion, we have pointed to the positive impact of oil on development, and how this can be seen in the resources directed towards development, as well as in the diversification of the economies and in the internal linkages resulting from oil-financed and oil-promoted development. We have likewise pointed to the serious failings of present Arab economic development, whether these have been directly created or intensified by the oil sector and the oil policies formulated, or indirectly brought about because of certain tendencies that the oil era has produced.

Whatever the nature and sources of the failings, a penetrating scrutiny of the broad horizon of oil policies would reveal that there is no policy that does not have its impact on the course and content, and therefore the outcome, of development. Beginning with exploration and production, through pricing and marketing, down to refining and the industrialisation of oil — all the policies, in one way or another, influence development. The most direct and visible influence is quantitative, in the sense that the reach and intensity of development reflect the volume of resources made available to the development effort. The less visible influence is social and attitudinal, to the extent that the availability and volume of oil revenues have a strong impact on the wants and demands of society, on the work ethic, on social cohesiveness, and on the breadth of the regional outlook in each of the oil countries.

To what extent these influences prove promotive of development, or distort its course and corrupt its content, is primarily the net outcome of the economic, political, and socio-psychological forces at play. Today, about one decade only after the take-over of the power to formulate and implement oil policies, the outcome is mixed. How it will turn out in the coming years is a matter that cannot be determined. But one thing is certain; it is largely within the power and the will of Arab society — leaderships and populations alike — to determine the nature of their future, either by correcting the failings of present development, or by permitting them to become more serious yet. If the concerned student of Arab development is to venture a guess, it would be that the implications of a disoriented and permissive attitude are serious and menacing enough to bring about the necessary corrections in the course of development before irreparable damage has been suffered. The 'period of grace' which the oil countries have before they translate their oil windfall into a renewable income at a high level is indeed short. And it is as non-renewable as oil is.

# 5

## OPPORTUNITY AND RESPONSIBILITY

### Introduction

The Arab oil policies of the 1970s have created impressive opportunities for the oil-exporting countries, but have simultaneously imposed heavy responsibilities on them. It is not possible to accept the opportunities and ignore or discard the responsibilities any more than it is possible to accept the responsibilities without assurance of continued opportunities. Furthermore, the responsibilities would not have arisen had it not been for the opportunities, and to continue to enjoy many of the latter requires acceptance of many of the former. This close dovetailing, which will become clearer as we proceed with the discussion in the present chapter, forces us to consider both sets together as we move along.

The opportunities, like the responsibilities, lie in the national, regional, and international arenas, although they are not identical in nature. To simplify the analysis, we shall examine oil policies and their effects in the national and regional contexts in the next section, leaving their examination within the international context to the last section of the chapter. In the course of the examination it will become clear that the enjoyment of the opportunities, like the shouldering of the responsibilities, can be effected only if certain appropriate and relevant conditions are satisfied. These conditions will be identifiable as we move along. But at this early stage we need to emphasise that a proper identification requires a proper assessment of the nature and magnitude of the opportunities, if overestimation or underestimation are not to distort the analysis and its conclusions. Likewise, a proper identification of the responsibilities requires a proper assessment not only of the true nature and magnitude of the opportunities, but also the assurance that those parties towards which responsibilities are felt acknowledge their responsibilities towards the oil countries in return and provide sufficient evidence of their willingness to discharge them.

Underlying both opportunities and responsibilities lie two fundamental realities. First, that the Arab oil countries are developing countries, struggling with the tasks of development like other Third World countries, and they are required to work hard and long to bring about the basic transformations in society and the economy that together constitute development. The second reality is that the relatively large financial resources which these countries have come to possess since the early 1970s are only large in comparison with the preceding phase of hardship. Furthermore, the source of these resources, namely oil, is a depleting and non-renewable resource, which could well be depleted before a developmental capability has been evolved, strong and pervasive enough to compensate for the loss of oil and to generate a national product capable of assuring the populations of the continuation of a reasonably-satisfactory level of living in the post-oil era. These qualifying realities are necessary to register in order for the magnitude of the opportunities and the responsibilities to be correctly assessed and not become inflated with fanciful expectations.

## The National and Regional Context

The *first* area in which the new oil policies have created important opportunities for the oil countries is obviously that of the acquisition of the necessary skills and institutions which the take-over of the power of control of the 1970s made possible. This area has been discussed at length throughout the book, but particularly in Chapters 2 and 3. What concerns us here specifically is the institutional part of the area: the expanded role of National Oil Companies, NOCs, under whatever name or designation they are recognised in the various oil countries.[1] The significance of this role is difficult to exaggerate, as the NOCs have taken over the functions of the foreign oil concessionary companies. Indeed, to the extent that certain NOCs handle downstream as well as upstream operations, they have a larger scope of activities within their jurisdiction.

The NOCs are in the process of acquiring a large body of capabilities and skills, of a technical, administrative, economic, financial, legal, and institutional nature. To begin with, their varied and weighty functions have necessitated the recruitment and organisation of large and highly-specialised work forces. But, more significantly, they have provided these work forces with experience, that most valuable form of education which can only be obtained through trial and error. The experience ranges widely over the areas of prospecting and exploration for oil, production, marketing, and pricing, as well as refining and petrochemical industries with respect to downstream operations. In addition to these far-reaching activities involving huge investments and very large transactions, there are ancillary activities which the NOCs have also to include in their area of concern, such as housing for staff, public utilities related to the upstream and downstream operations handled, and the building of infrastructure required for efficient functioning.

While it is true that the actual execution of much of the work involved is undertaken by specialised ministries or agencies, or by private contractors, the conceptualisation and planning, like the design and supervision of the bulk of the work, falls within the area of activity of NOCs. To cope with all the tasks involved, the NOCs scout thoroughly for talent in the oil countries themselves, but they also make substantial recruitment in the non-oil Arab countries, and to a lesser extent in non-Arab countries as well for highly-specialised jobs. The opportunities thus provided to the nationals of the oil-exporting countries (and in part to other Arabs) are a most valuable form of education. But one aspect of the educational experience is essentially restricted to the nationals:

decision-taking in a wide range of areas on weighty matters involving substantial issues and financial resources, and of most significant and far-reaching implications to the countries involved, to the region, and to the world at large.

The institutional structure empowered to act at this level of decision-making naturally goes beyond the NOCs narrowly identified and defined, to include not only the relevant ministers (of oil, finance, national economy, planning, and often foreign affairs), but the government as a body, and even the head of state. None the less, the NOCs have a sensitive part to perform even though decision-taking is made at a higher rung of power: this is the preparation of the case for the particular decisions proposed, whether they relate to the volume of production, the price of export crude, or the refining capacity to be developed.

In a directly-related sense, the magnitude of the role of the institutions concerned with the operation and management of the oil sector broadly defined, and therefore the magnitude of the opportunity this affords to the oil countries, is mirrored in the magnitude of the responsibility it entails. We refer here to the implications of the decisions taken for the volume of oil (and gas) put on the international market, for the prices charged, for the earnings of the oil exporters and therefore for the transfer of resources between oil exporters and importers, for international trade in non-oil categories, and finally for the economic growth of the oil consumers. Here, as in the many other instances to be encountered in this chapter, the interdependence of opportunity and responsibility and the interdependence of producers and consumers, are affirmed. And here, likewise, the significance of the establishment of *true* interdependence is affirmed – an interdependence which does not worsen the already existing imbalance between the developed and the developing worlds but corrects it. This can be achieved only if the opportunity afforded by decision-making is translated into meaningful development, and the international economic order is so adjusted and corrected as to narrow the gap between the powerful and developed and the weak and underdeveloped, as we shall see later.

The *second* aspect of opportunity which the new policies have made possible is greater solidarity and cohesiveness among the oil-producing countries — and this applies to all 12 of them, not just the seven that are members of OPEC. The solidarity relates narrowly to oil policies and activities. We indicated earlier that the oil governments have substituted for the tight structure of decision-making and operation formerly built by the foreign oil companies, one of their own; but we also indicated the ways in which this new structure differs from the one replaced. Part of the difference points to the superiority of the older structure, to the extent that the cohesiveness of the companies with respect to the price charged, the volume of oil produced, and the markets supplied was easier to come by and more binding on the company group. (This was particularly true of the period when the seven majors alone controlled the industry.) Another part points to a potential, if not altogether realised, advantage to the governments, to the extent that they now undertake joint programmes and activities within OAPEC,[2] and to the further extent that some OAPEC members have undertaken the establishment of some important downstream projects together.

The opportunity to formulate varied oil policies together, followed by the desire to do so and the actual movement in that direction, are significant developments for the individual countries concerned, and for the Arab region as a whole. The rationale behind this statement is obvious, but it must be further affirmed by the contribution that the development makes for the general cause of economic integration generally, beyond the oil sector. We do not mean that Arab economic integration is taking place on a scale and with an intensity meaningful and satisfactory enough to Arab nationalists. But a first phase is being successfully passed through, namely that of co-ordinated co-operation. For the countries concerned to be promoted from this first phase to that of far-reaching integration would require much greater mutual political confidence and affinity, a much stronger determination to integrate, but above all a genuine conversion to the belief that integration brings net beneficial results to the parties to it, that is to individual countries, in addition to the benefits it brings to the region at large and to the oil industry in particular.

Such benefits are demonstrable theoretically and analytically, but they must be demonstrated in real experience, on a large enough scale, to influence political attitudes and shape decisions duly. No doubt certain vested interests will fear that they will suffer through integration, and many politicians will feel that their own oil industry will suffer if downstream operations were integrated, whether horizontally or vertically, with those of other oil countries. Here lies important responsibility that must be discharged, one of public education and information. Politicians, economists in public and private life, organised bodies, and the media must all join forces to bring about a better understanding of the case for integration.

But the pursuit of solidarity among the oil countries also lays a responsibility on the major oil-importing countries — that is, the industrial countries. The desire for solidarity, policy co-ordination, and a large measure of horizontal and vertical integration among the oil-exporting countries must be correctly understood, even when and where it influences the determination of the price of crude, the volume of production, or where it expresses itself in concern with the purposes for which the crude is utilised. To begin with, there is a need for the consumers to realise that producer solidarity is not pursued in a spirit of confrontation and conflict. It is rather motivated both by legitimate self-interest and concern for the optimisation of the development and export of the scarce and most essential oil resource; and by a sense of international responsibility to the extent that the volume of oil consumed and the nature of its use cannot but be matters of great import to the producers and the consumers alike, if a global view of energy and the finiteness of oil are to be kept in mind. The wise consumption of oil and gas, by the whole world but especially by the large industrial consumers, cannot be a matter of indifference to any country, least of all the large consumers; nor, obviously, can it be a matter of indifference to the producers.

If this argument is sound, and we submit it is, then the solidarity pursued by the producers is in effect not merely internal solidarity among themselves, but also real solidarity *with* the consumers. Or, phrased differently, there is here a clear case of interdependence. But, once again, to offer such an interpretation of interdependence places the responsibility of proper response on the consuming industrial countries. The interpretation offered should solicit a response of understanding and true co-operation. Unfortunately, as indicated earlier in the last chapter, often there have been cries by politicians and economists in certain industrial Western countries that OPEC must be condemned for its pricing policies.[3]

Yet, in fairness, it must be admitted that there has been increasing understanding of the intent and policies of OPEC, and positive interpretations have been placed on these. Greater credit in this connection must go to those politicians and analysts who took such an attitude all through, even in the stormy days of 1973/4. Finally, conversion to the cause of interdependence and pursuit of a course leading to it, can only be possible and genuine if the two interdependent parties are not very far apart in the scales of power. They certainly are now, and will continue to be for a long time yet. However, the imbalance can be substantially corrected if the industrial countries were to realise how much in their interest it is to help the oil countries accelerate their national and regional development, and that to achieve this, these latter countries must achieve solidarity around their basic oil policies.

The basic oil policies around which there is a valuable opportunity for solidarity are rather clear and relatively easy to define and formulate. They run the whole gamut, from exploration on to the most distant petrochemical industry downwards. Yet they are very difficult to place together within a delicately-balanced system which is designed to operate efficiently, smoothly, and with rigorous internal consistency. To design such a system is both a challenge and an opportunity, and also a weighty responsibility. And this is the *third* aspect of the opportunity/responsibility relationship which is being addressed in the present chapter.

Such a finely-tuned system in operation would attempt to achieve some seemingly-contradictory objectives, all at the same time. Thus, it would stipulate for active exploration, and for improved techniques of recovery, in order both to assure the producers of larger reserves, and also to assure the consumers that oil resources can last longer. At the same time, the system would stipulate for responsible, disciplined production policies, even though reserves may now increase faster than before. This discipline would be called for, once again, by the interests of producers and consumers alike. For the former, it would mean a less permissive and more discriminating revenue spending policy, the need for a lower (or a stable) level of production, and a longer life span for oil. For the consumers, this discipline would mean some stringency in the availability of oil, therefore greater discipline and discrimination in consumption, and a more active search for alternative sources of energy leading – among other things – to a reduction in the pressure on oil and the preservation of more of it for more essential purposes like fertilisers and petrochemical products, as well as those uses as fuel for which there is no substitute.

In line with the considerations underlying the suggested aspects of the system would be the imposition, by the oil-producing countries, of greater discipline in energy consumption domestically and in the Arab region as a whole. This would be called for in order that the disturbing estimates of high consumption be qualified considerably in practice. These estimates exceed half the estimated oil production of the whole Arab region by the turn of the twenty-first century.[4] To impose such discipline, through pricing, better information, and other measures within the Arab region, would be consistent with the insistence that the rest of the world, particularly the industrial countries, should do the same. Pricing is a central factor in the kind of system under discussion. A gently-rising tendency would be necessary in order to provide the logical and practical underpinning for discipline in consumption, both within the oil-producing countries and the oil-importing countries. It would also be necessary, in order to provide a greater motive and justification for a more active search for substitute sources of energy which are within a competitive range with oil, in terms of price, as well as quantity and quality.

As far as oil producers are concerned, pricing and production policy has to take account of additional considerations. One of these is the need to have a certain minimal volume of associated gas for some types of essential use, particularly in the Arabian Gulf region, such as desalination, electricity-generation, and the provision of feedstock to petrochemical industries. Furthermore, in view of the fact that energy consumption in the producing countries is still low owing mainly to the low level of economic development, the price mechanism as applied domestically should permit for some expansion in consumption. However, the present permissiveness in consumption should be greatly curbed. If it is not, the Arab region would find itself with considerably reduced quantities of oil to export, what with its own consumption plus that of other Third World countries. On the other hand, there would be an irreducible minimum of revenues which it would be necessary to earn from export, and here the factor of revenues would play its important part in influencing the volume of production.

The approximating process which would be necessary to bring about equilibrium in this system would simultaneously involve the following factors: activated exploration and improved recovery in order to reduce the level of anxiety and uncertainty over the future of oil reserves; a more or less steady level of production, with a ceiling of, say, 18-20 million barrels a day for the Arab countries as a whole, in order not to impose crippling cuts on the world economy and yet to permit a reserves-to-production ratio which is higher than would otherwise prevail were the ceiling to be substantially higher (as it was, for instance, during the 1970s); enforced discipline on consumption, in order to promote sound consumption habits among the producers, and to reduce the pressure on their production and simultaneously allow for greater exports; a gently-rising price level in order to cope with a gentle level of inflation and to benefit from a part of the economic growth in the industrial countries which the oil resource helps to bring into being; permitting the producers to have the associated gas and the oil they need for their downstream operations; and maintaining revenues at their real level and thus providing the finances needed by the oil producers.

The final factor involved in the equilibrating process requires that the producing countries should use much greater discipline and foresight in their spending, whether it is for consumption or investment, since considerable savings are called for and possible in both categories. Any savings made will reduce the internal pressure for larger production than stipulated in the system, and therefore would remove the danger of disruption of the system. Obviously, the nature of the relationships described strongly suggests that they have national, regional, and international aspects. They obviously involve mutual understanding and mutual accommodation, in a context of true interdependence. Finally, it is perhaps clear by now that such a system is finely-tuned in the sense that the behaviour of each of its components influences and is influenced by the behaviour of the other components. And the national, regional, and international policies relating to energy are of direct relevance to it, as it would be of relevance to them. Consequently, a smoothly-operating energy relationship between the oil-exporting and the oil-importing countries would have to preserve the integrity of the system and the smoothness of its operation. If this requirement is met, then the opportunity for a healthy and mutually-beneficial interdependence would be met, and the result would be a constructive interaction between opportunity and responsibility among oil exporters and oil importers.

The *fourth* and final aspect of the opportunity/responsibility relationship to be considered here is that of national and regional development and security. We have had occasion to refer to development in this chapter in different contexts.

Therefore all that is needed at this point is to emphasise the enormous developmental opportunity the new oil policies provide, and the responsibility that attaches to the opportunity, both for designing and seeking healthy and comprehensive development, and for guarding against false objectives and courses of action which would abort the desire for development and would lead to the dissipation of resources presumably directed to it.

In fact, a careful look at the role of the new oil policies in satisfying national and regional needs would reveal that oil provides not just the possibility for the pursuit of development, but also for the satisfaction of certain consumption needs that were pressing before oil resources made attention to them possible; likewise, the resources have provided the means for the assurance of a better capability for safeguarding national and regional security, broadly defined as we suggested in the last chapter. The reason, therefore, that we concentrated on development alone in the preceding paragraph was that development in the long run is the real fountainhead for the capability to provide both for consumption and security needs. Such a capability cannot be maintained without real progress in the drive for development. In other words, oil revenues by themselves are bound to dry up one day, if not replaced by and translated into developmental capability which would ensure the continuous generation of national income at a satisfactory level.

At the national level, that is, within the context of individual oil-exporting countries, the opportunity for development is vast, considering the financial resources available since 1974. But the magnitude of the opportunity is illusory, if we do not insist that development should mean national capability for a high-level economic performance, with all that goes with this performance in institutional, technological, socio-cultural, and political transformations, at the public level, among social groupings, and down to the individual level. So far, the extent and content of these non-economic transformations falls considerably short of the economic transformation as witnessed in economic and social infrastructure, building and construction, and the promotion of refining and petrochemical industry — to name the leading areas of achievement. However, this achievement in itself also far outdistances the productive capability of the indigenous labour force in each of the countries concerned — that is, it is beyond the design, construction, and managerial and operational ability of the national manpower.

This series of gaps between what appears on the surface and what lies underneath, and between national capability and imported capability, qualifies the performance considerably. It indicates that the promise of the developmental opportunity provided will still need many years before it can be fulfilled, through an effective marriage between financial resources on the one hand, and, on the other, manpower, scientific and technological abilities, organisational forms, institutions, and the vision and will that discipline and order the other requisites. This does not mean that the opportunity has been lost; far from it. It means that to seize such a historical opportunity and turn it into concrete realisation calls for a different set of developmental visions, policies, institutions, capabilities, and — above all — time horizons than it has been possible to have since the early 1970s.

This discrepancy draws the contours of the responsibility that lies on the shoulders of every social group in each of the countries, including academic and intellectual leaderships, business leaderships, labour leaderships and other syndical groups, public opinion leaderships, in addition to political leaderships. Given the innate intelligence and dynamism that the Arabs have shown themselves to possess since their early history, it is reasonable to expect that they will not be very slow in displaying the ability to correct the errors that have beset the course, the structure, and the content of development so far pursued. The crucial point is that correction needs to be effected before a great deal of time and excessive efforts and resources have been dissipated. The hopes of millions of citizens will suffer considerable and dangerous frustration once they discover that their reward is twisted and marginal development, with growing — rather than dwindling —

dependence on the industrial world, which itself had been principally instrumental in the economic under-development which it is now hoped to abolish.

Important as the satisfaction of national needs is, it must be supplemented and complemented with that of Arab regional needs, again for development and for security. Here again, as indicated in the last chapter, the performance is even less creditable than at the national level. Attending to the regional needs is not an act of brotherly compassion, although there is a strong element of this compassion involved. It is more an act of national self-interest, owing to the strong positive effect of the interaction between national and regional development (and security) efforts. This is basically because, as indicated earlier, the Arab region constitutes the strategic economic (and security) depth for each of its constituent parts – in terms of market, supply of manpower and material resources, scope for undertakings the scale of which require large areas and populations, the establishment of a strong economic base for security, or the establishment of a strong security belt to protect economic resources and achievements.

The opportunity that the oil-exporting countries have to help the rest of the region develop is great, but so is the opportunity for the rest of the region to help the oil countries develop. The two parts of the region are complementary, given their different human and natural endowments. Awareness of this interdependence exists, but not of the magnitude of the true opportunity. Furthermore, supportive developmental action falls short of the magnitude of the awareness, limited as this is. Here lies weighty responsibility on both parties to the interdependence: the oil-exporting countries for making their support match their declared awareness, and the non-oil countries for making the developmental opportunities and investment openings better known to the first group and, simultaneously, making the flow of resources to these openings more rational (because preceded by and based on better studies), more secure, and therefore more attractive. The discharging of their share of the responsibility by both groups of countries would see more of the financial resources of the oil countries, which are now in exile in Western markets, flow back to the Arab region.

It ought to be added that attention to the regional dimension calls for more than financial flows from the oil countries to the non-oil countries, or for skilled manpower flows in the opposite direction. It calls for many programmes and projects that are regional in objective, scope, design, financing, execution, and finally operation. Three prominent areas of action to point out in this respect are the development of scientific, technological, research, and training programmes meant to upgrade Arab manpower and to use the available capabilities more extensively and intensively; the development of the vast potential for agricultural (and food) production that is grossly underutilised now. The present rate of skilled manpower utilisation leaves a great deal to be desired.[5] This may sound paradoxical and may therefore raise some questioning, considering the widely-proclaimed shortage of skills. Yet it is a fact, especially as far as high-level manpower is concerned, where a great deal of misallocation is in evidence. The paradox of agricultural potential and simultaneously of serious exposure of food security is no less striking.

The serious implementation of the Strategy for Joint Arab Economic Action, and the Charter and National Plan which flow from it, would go a very long way towards the correction of the course of national and regional development, the enriching of the flows between them, the expansion of the Joint Arab Economic Sector, and finally the increase in the developmental rewards made possible through oil and a better sharing within the Arab region as a whole. It must be added that this would not mean that the non-oil countries would claim a proprietary right on the oil resources of the oil countries, but that the latter would divert a larger portion of their financial resources which are now in Western money markets into productive investment in the Arab region (and into investment elsewhere in the Third World).

This would be one way of protecting these resources from the considerable erosion in their value which they now suffer because of inflation. To turn bank accounts and other paper (liquid) investments or placements into real productive investment is demonstrably an effective safeguard against inflation. On the other hand, the non-oil countries have to exercise much greater discipline in curbing waste, adopting sound and firm economic policies, marshalling a much larger volume of domestic resources, and establishing stabler economic and social systems — if they are to succeed in attracting substantially larger Arab investment funds.

A national-cum-regional response to the opportunity/responsibility relationship, which aims at benefiting from the new oil policies to achieve solid and meaningful development, and security in its fuller sense, is not meant to be the course to pursue merely in order to optimise present development. It would also be the optimal course to prepare for the post-oil era, when oil gets to be considerably depleted, or downgraded in the scale of energy resource availability. The Arab oil-exporting countries have a unique opportunity in their hands and ahead of them, but it lays a very heavy responsibility on them. And, to the extent that the rest of the Arab region wishes to benefit from the opportunity, it also has to shoulder its share of the responsibility. In other words, the Arab region as a whole has yet to prove its proper appreciation of the magnitude of the opportunity, by proving its ability to assess and discharge its share of the responsibility.

### The International Context

There have been repeated occasions when the discussion of oil policies in their national and regional context has had to move into the area of international relations. This is unavoidable because of the close interrelationship between the two areas. Otherwise, the main focus and emphasis in the last section has been on issues more particularly pertinent to the national and regional framework of policies.

The opportunity/responsibility relationship to be discussed

within the international framework will be looked at from three angles. The *first* is that of the nature and style of the relationship between the oil-exporting countries on the one hand, and the developed industrial countries on the other. By the second group we essentially mean the Western or OECD countries, as they are the major partners of the first group with respect to trade in oil and other goods and services, both as far as imports and exports are concerned. The basic question here is: What will the relationship essentially be, one of co-operation based on the premise of interdependence, or one of confrontation based on the premise of the irreconcilability of interests? And, if it is going to be one of co-operation, as so far we have argued that it could be, and should be, will the interdependence underlying it be a mere euphemism for highly unbalanced relations meant to lull the oil-exporting countries into complacency, or will it be a concrete reality which provides scope for a fair exchange of interests and mutual respect?

Several issues suggest themselves as pertinent to this enquiry. They include the price at which oil will be supplied: its level, its direction, the stability of its course (that is, will it maintain a rather even course, whether rising gently, falling gently, or remaining unchanged, or will it fluctuate sharply?); the supplies put on the market — their absolute size, and their stability; the extent and seriousness of the penetration of the oil-exporting countries into such areas of activity as transport of oil and gas by tankers, refining, petrochemical production, and distribution overseas; and the attitude of the advanced countries to all these matters and the intensity and style of their response to the action of the oil suppliers in the relevant areas. We have had occasion to refer to all the issues raised above. The price issue in particular has received considerable attention and space in this book. Therefore we will only consider one issue here: that of supply.

The politicians of OECD countries, like much of the literature emanating from their region, never tire of repeating that they are basically and fundamentally concerned with the certainty or security of supply of oil, given its central place in their economies and its vital role in their growth and security. This concern is fair enough, and neither the oil-exporting governments nor the economists and analysts in their countries ignore its centrality or its legitimacy. Furthermore, this concern is usually coupled with the stipulation that the oil should be 'reasonably' priced — meaning, that the price should be moderately if at all rising and generally stable and predictable.

Let us call this composite desideratum of volume plus price the 'security of supply', for brevity. Accepting this desideratum as fair and legitimate is of course tied to the volume of supply which is to be made certain. At this point we are forced back on the volume of production which we examined earlier, in the national and regional context of oil policies. In other words, the consuming countries will have largely to accept as given that volume of exports which the producers are able and willing to put on the market, if this volume has been determined in the light of the complex of factors that we discussed earlier.

It will be recalled that these factors included the national and regional needs of the producers and the revenue they call for, therefore the price that is consistent with the volume; the consumption requirements of the producers and their Arab region; the size and growth (positive or negative) of oil reserves; and the policy decision regarding the life-span desired for the reserves. The residue will be available for export, and it will have to be apportioned between Third World and OECD countries (assuming the socialist countries to be marginal importers from the OPEC or OAPEC countries).

The ratio between these two groups will depend on the availability of non-OPEC oil and non-oil energy to each country, and on ability to pay the import bill. (With respect to the latter point, the oil exporters will have to continue to help the needier Third World countries to face their balance of payments hardships as they have done since 1974). Whatever the volume available for export, it is essential for exporters and importers alike that this volume should have a high degree of certainty about it. This could be achieved through appropriate contractual relations or through mutually accepted implicit patterns of trade relations.

But the security of supply of oil is one side of the coin of international relations. On the other side of the commitment, there will have to be security of supply of capital goods and technological help from OECD to oil-exporting and other Arab countries, at reasonable prices, to enable them to build up their production and development capability as speedily and as firmly as possible. Obviously, to a large extent it is up to the oil exporters to adopt those policies which would enable them to move towards meaningful and self-reliant development. It would be less than honest to maintain that they have exhausted their possibilities in this respect, and that their failings are all to be blamed on the industrial world. But it would be equally less than honest to pretend that the industrial world is offering its help in earnest and at reasonable prices, or that it is endeavouring to enable the Arab countries to acquire true developmental capability, essentially in the area of appropriate technology. If anything, the experience of the past seven or eight years since 1974 shows that the kind of 'help' extended has tended to lengthen and intensify the dependence of the Arab on the industrial countries.

This writer feels that the blame should be shared by both groups, and that the burden of correction should mainly lie on the Arab countries, in the sense that they should chart a new course of action and move along it with determination, as a necessary condition to make the industrial countries change course. But the necessary condition needs a response to become both necessary and sufficient: the honest willingness of the industrial countries to meet the requirements of the Arab countries, with the quantity and quality of help most conducive to the true acquisition by the latter of technological capability. It is true that the process of acquisition calls for an earnest recourse to one's abilities and resources, an intensive resort to learning-by-doing and trial and error, and an active programme of scientific and technological development at home. But this does not absolve the industrial countries of the charge of deliberate failure to shape and direct their help so as to harmonise with the essentials just listed as falling in the area of necessary action by the Arabs.

The security of supply also depends on the satisfaction of other conditions or circumstances. These are obvious and will only need to be listed to be recognised. As indicated earlier, first and foremost, they depend on the oil exporters' perception of their own, and general Arab and Third World needs, translated into a certain magnitude of revenues. They also depend on the exporters' perception of the needs of the industrial countries, and this in turn is influenced by the profile of uses to which oil is put in the latter group of countries. The relevance of this last consideration arises from the growing tightness of the hydrocarbon resource, and the desire of its owners to have it stretch as long as possible into the future. Another factor of relevance to the volume and security of supply is OPEC's politico-economic relations with the industrial countries, and the extent to which co-operation or confrontation is the prevailing mood. Yet another factor is the progress made towards the correction of the existing international economic order, and the attitude of the industrial countries to the pressure by the Third World for a New International Economic Order, NIEO.

This NIEO, it will be recalled, is desired because it is vested with a great deal of hope in the Third World, as in certain quarters in the industrial world itself: hope that it would be more efficient because less wasteful of resources; more flexible and fair in the redesign of a new international division of labour, under which developing countries would have a greater scope for industrialisation and the export of manufactures; more just in the distribution of the benefits of the international system, in the sense that the developing countries would get a fairer return for their raw materials, and be charged less exorbitant prices for their imports, particularly those of capital goods and technical services; and more accommodating to the developing countries by allowing them a greater say in international economic organisations and in the design and conduct of global economic conceptualisations and policies.

The oil-exporting countries believe, and in this they are supported by other Third World countries, that their oil puts in their hands a card which should permit them admission to the world's highest economic councils, there to talk for the Third World as a whole and to argue its causes. In the final analysis, it is only because of oil that, since 1974, the Third World has had some countervailing power in its hands, with the capability of partial deterrence in the face of the economic omnipotence of the industrial world. Yet the power and authority that the oil-exporting countries have acquired in world economic councils lays on them an enormous responsibility. This is that they should keep their close identification and solidarity with the Third World and its causes. So far they have not betrayed this trust.

The last factor to list here is specific to the oil-exporting countries. This is that the security of supply to a considerable extent depends on the attitude of the industrial countries towards the insistence by the former group that, for the supply to be secure, and to be maintained at a satisfactory level, there must be considerable quantities of oil in the earth to count on. For this to be assured, the oil countries must have much more exploration activity undertaken, and they must have advanced methods of recovery applied, in order for many more oil reservoirs to be tapped, and for those already tapped to be more effectively exploited. The operations involved are highly technical, and above all very expensive. If the major beneficiaries of increased oil supplies want the potential to be turned into actual supplies, along with security of supply, they must be ready to contribute with their expertise as well as their financial resources. Otherwise the oil countries would feel less justified in undertaking much more exploration and costlier methods of recovery, if the benefit but not the costs are to be passed on to the much richer and more advanced industrial countries.

Important as all these factors are for the assurance of certainty of supply which the industrial world seeks, and important as the considerations are which the oil producers insist on introducing as inputs in their response and policy elaboration, they all remain insufficient to establish a balanced and healthy relationship between oil exporters and oil importers — particularly the major industrial countries among the latter. For, while these have a legitimate expectation of security of supply, the oil exporters have an equally legitimate expectation of *security of demand*. This expectation is all the more pressing now, with the conservation measures and practices taken by the large consumer countries, and with the considerable substitution taking place. The substitution is first between oil and coal (and to a much lesser extent other forms of energy), and secondly between OPEC oil and non-OPEC oil. OPEC oil is left as the residual energy to resort to, which makes it extremely difficult for the producers to plan their budgeting, given the possibility of wide fluctuation in exports and therefore in revenues. The implications of unpredictability for development, security, and consumption are so enormous as to be intolerable.

If it is fair that the consuming countries should be able to count on a secure volume of oil imports, it is equally fair that the producing countries should be able to count on a secure volume of oil exports. Indeed, it is more vital for the oil exporters to have such security, given the fact that oil is central and crucial to their consumption, development and security, as well as to their gross domestic product, foreign exchange earnings, and fiscal revenues. While oil is also important to the industrial countries, it is neither as central nor as crucial, nor does it occupy a place anywhere as high in any significant economic indicator. OPEC (and OAPEC) oil must not be a residual supply, nor should demand for it remain a residual demand, if interdependence is to be genuine and not merely a play on words.

The *second* aspect of the opportunity/responsibility relationship relating to oil in its international context is the political leverage that oil bestows — or is supposed to bestow — on oil producers.[6] Nothing need be said to camouflage or sugar-coat the fact that possession of as vital and strategic, and now as scarce a resource as oil, could provide its owners with certain political leverage in the world. This should be no cause for jubilation among the oil producers, or for condemnation among the oil importers. It is a fact of political life.

But it is a fact the assumed implications of which are far above the actual implications. In other words, the Arab oil exporters have utilised oil as a political (or politico-economic) pressure mechanism only very sparingly, and when they did utilise it thus, they did so with insufficient stamina. We refer here to the cuts in production, and the selective export embargo, resorted to in mid-October 1973, on the occasion of the Arab-Israeli war. This utilisation, it will be recalled, was to register extreme displeasure and anxiety at the support extended to Israel, and to mobilise international opinion and have it translated into pressure on the United States, which was supplying Israel with massive *matériel* to enable it to hold on to occupied Arab territory. It was hoped that the oil sanctions would thus make the United States, in turn, put pressure on Israel to make it withdraw from those Arab territories which it had occupied in the war of June 1967 — namely, the West Bank, the Gaza Strip, and East Jerusalem in Palestine, Sinai in Egypt, and the Golan Heights in Syria.

We do not want to go into any detailed discussion of the use of oil as an instrument of pressure with respect to the Palestine cause, but merely to make three specific observations. The first of these is that in a vital struggle involving life or death to a society, the use of all resources — human and material — in the struggle is resorted to; this is universally accepted as a principle and applied in practice. Oil is a resource that was, and continues to be considered powerful, and it would not be understandable to abstain from using it.

The second observation qualifies the first. It is that the use of oil as an instrument of political pressure is not a normal part of Arab oil policies, but one which is reluctantly considered and utilised as a last resort. The Arabs look at their oil as an economic resource, essentially meant to provide them with developmental capability and with defence capability to protect their rights and interests. This oil achieves through financing development and service as its engine through being a leading sector, and — as far as security is concerned —

through financing arms purchases and defence generally, both nationally and regionally. But it is not viewed as a political resource to keep at the ready for political pressure. This is a position of principle and of practicality alike.[7] The Arabs are fully aware of the question of practicality, considering the fact that the Western countries have counter-instruments of pressure in their hands that they could, and would, use if the Arabs used the oil instrument. We refer here to Western exports of food, technology, and arms – all of which are as vital to the Arabs as oil is to the Western world. And, in addition, the West has (as of the time of writing) over $350 billion of Arab financial reserves which stand to be frozen by the United States and possibly some other Western countries, were the oil pressure-mechanism to be used again. While it is true that some protection can be designed and obtained against the use of the counter-instruments of pressure, it is also evident that this would only be possible in the medium term (two or three to five years), and the Arab governments do not seem to be in the process of taking such precautions.[8]

The third observation in the present context is that the Arabs have not invented the use of economic resources or capabilities for political purposes. This is an invention as ancient as world strife. But the sceptical reader may need to be reminded of the economic pressures applied in modern times: the Western economic sanctions and pressures put on the Bolshevik revolution in its early days, the oil embargo applied by the United States, Britain, and France against Italy when it invaded Abyssinia in the mid 1930s, the protracted economic embargo of a long list of goods by the United States against China, the USSR, and Cuba, since the 1950s to the present. The Arabs feel that the cause for which they applied their oil deterrent and/or sanctions is more vital to their existence than the causes which variously moved Britain, France, and the United States into action were to these Western countries.

Another aspect of the political leverage which oil puts in the hands of Arab exporters is the improvement in their negotiation and bargaining position in various international organisations and forums. These include the World Bank and the International Monetary Fund, the United Nations and its various specialised agencies like UNIDO, UNCTAD, ILO – to name only a few – as well as in the so-called North-South Dialogue and the meetings and deliberations to shape the NIEO. The content of the leverage involves many issues, such as the transfer of more resources to the poorer countries for development and for balance-of-payments support, the encouragement of industrialisation in Third World countries, the solution of the problem of huge indebtedness by these countries, the supervision, control, and correction of the excesses and malpractices of transnational corporations, and like issues that have heavily burdened Third World countries and darkened their economic horizons.

The oil-exporting countries, though richer in terms of financial resources, are subject to much the same concerns and stresses as other developing countries. The fact that they have the oil resource, and that this gives them some leverage, provides them with an opportunity to work on their own behalf and on that of other non-oil countries. It also places on them the responsibility to work sincerely and consistently, and not to allow themselves to be diverted by their transient financial advantage and the misalliances it might trap them into. Only if the developing countries have a clear and thorough understanding of their situation – whether or not they have oil – and exercise steadfastness and wisdom in their demands, will the industrial countries accept the introduction of the long overdue corrections into the international economic system.

The *third* side of the opportunity/responsibility relationship is that between the oil-exporting countries and the oil-importing developing countries. One aspect of this relationship has already been referred to, though briefly, in the last paragraph. Three other aspects need to be looked into here. The first is immediate: it is the urgency of having the weight of the oil import bill lightened by help from the oil-exporting countries. This is being undertaken, partly by aid from the OPEC Fund for International Development, OFID, but also through bilateral official aid. Although much of the aid is directed to development, it has the effect of facilitating the payment of the oil bill, since the credits and grants extended increase the availability of resources to the recipient countries. However, not all oil-importing developing countries need aid, as many among them (and they are large oil consumers) are in good economic condition. These include the faster-industrialising countries in South-East Asia and Latin America.

The second aspect is the possibility, and the initiation, of programmes instituted by the oil countries to help the developing oil-importing countries explore for energy. This form of aid is of a more lasting nature than the first direct aid form. The help can be financial or in kind; it would be best to come in both forms. The main constraint at the present is the limited capability of the oil exporting countries to undertake prospecting and exploration activities for oil and gas, with all the technical operations, and the ancillary works, involved. But as the capability increases, the scope for the form of aid under discussion expands. And, while this would mean considerable help to the non-oil countries, it would also be of value to the oil exporters themselves. It would provide them with more room for gaining experience, widen the geographical and technical scope of their activities, and provide a larger opportunity for their investments. But given this opportunity, the oil countries must guard against behaving in an exploitative manner, or in such a way as to be considered and felt as alien as the foreign oil companies were considered and felt to be in the Arab countries before the 1970s.[9] Here lies a challenge and a responsibility for the Arab countries to design a pattern of relationship which would be truly co-operative, while bringing in a reasonable return on the efforts and investments made.

Having said this, it remains necessary to add that the help to be extended in the area of exploration must not, and indeed cannot, all or mostly be provided by the Arab oil exporters or even by OPEC as a whole. The industrial world has to bear the major share of the burden, both financially and technically. The special facility that it is hoped the World Bank will establish must be adequately financed and supported by the industrial countries. In fact, this would be one other aspect of interdependence, since the success in finding oil and gas in non-oil countries would increase the global supply of hydrocarbons, delay the crunch of serious oil shortages, provide more time for the search for and development of competitive alternative sources of energy, and generally contribute to economic growth and development.

The support of Western countries may well come, in part, from private companies with experience in the field. This has obvious advantages. But it has serious potential disadvantages that can be identified from the long experience of the oil countries during the foreign company regime. Consequently, in addition to the limitation of the role of foreign companies (and its design in the form of service contracts), the World Bank should play the major role in helping non-oil developing countries explore for oil, thus minimising the danger of exploitation by transnational corporations placed face to face with poor and rather resourceless countries.

The last aspect to look into is foreign aid by the Arab oil-exporting countries to needy Third World countries. The present writer hopes it would not be presumptuous and self-congratulatory for an Arab writer to stress the commendable record of the Arab governments in the field of aid. This record is not only good, but far better than that of OECD countries, if aid is to be measured as a proportion of GNP. Furthermore, it is not by any means restricted to the Arab countries, but covers countries in the whole Third World. Indeed, the 1981 OECD report on development assistance[10] indicates that bilateral OPEC aid going to non-Arab countries rose from 16 per cent of the total in 1979 to 20 per cent in 1980, but that aid through multilateral Arab and OPEC sources rose from 41.5 per cent of the total to 45.5 per cent for the two years respectively.

Concessional assistance by Arab OPEC members for the period 1970 through 1980 totalled $37,577 million according to the report referred to, whereas aid by the non-Arab members (namely Iran, Nigeria, and Venezuela) totalled $3,160 million for the same 11 years. In fact, this last group extended no assistance whatsoever before 1973, and gave a mere $25 million for that year, against $1,283 million by the Arab donors. Average Arab concessional assistance for the seven years 1974-80 (which reflected the full impact of price and revenue correction) amounted to $5,009 million a year. Arab assistance during the period 1970-80 amounts to a sizeable proportion of GNP; it ranges between a low of 2.56 per cent in 1979, and a high of 4.99 per cent in 1975. Even during the years 1970, 1971 and 1972, before the price adjustments of 1973/4, Arab concessional assistance was 4.04, 2,94 and 3.18 per cent of GNP for the three years respectively.[11]

These proportions of aid are far superior to those of OECD countries which average about 0.33 of one per cent of aggregate GNP. Furthermore, in order for the reader to realise that no statistical trick is being played here, we add that, in absolute terms, aid by all OECD countries together totalled $22,267 million for 1979 and $26,603 million for 1980, and averaged $18,699 million a year over the six years 1975-80. During these years, and in percentage points of GNP, official assistance from OECD countries ranged from a low of 0.09 (for Italy in 1979) to a high of 0.99 (for Sweden in 1977).

It is noteworthy that Sweden, Norway and the Netherlands have by far the most creditable record in percentage points, but none ever reached a full 1 per cent. The least creditable record is that of the United States (average 0.27 per cent over the period 1975-80), Japan (0.24), Switzerland (0.20), Finland (0.19), and Italy (0.12). It is both interesting and saddening that the group of lowest donors includes three very opulent societies: the United States, Japan, and Switzerland, and that their low contributions are not an accidental deviation from a good record, but in conformity with a consistently poor record ever since 1970.[12] To drive the point further home, and then leave it to speak for itself, we indicate here that the aggregate GDP of the Arab donor countries referred to here was $209.6 billion and $318.8 billion for 1979 and 1980 respectively,[13] whereas the GNP of Canada alone was $222.5 billion, and that of Italy $298.2 billion, for 1979. (The data for 1980 are not available in the source.) Together, the OECD countries enjoyed an aggregate GNP of $6,317 billion for 1979,[14] just over 30 times that of the Arab oil-exporting countries. On the other hand, the development assistance extended by the latter group averaged about 3.39 per cent of their aggregate GNP — that is, a proportion which is 10 times larger than that for the OECD countries for the same six years 1975-80. (These are simple not weighted averages, but they are good enough for present purposes, as general orders of magnitude. A weighted average would show a wider discrepancy, owing to the fact that the United States has the largest GNP in OECD, but makes a very small percentage contribution.)

Finally, it remains to be added that the OPEC countries are themselves developing countries still very far from a state of satisfactory development, and that they are giving aid out of the proceeds of the sale of a depleting, non-renewable resource, while the OECD countries extend aid from a developed economic base and a self-generating income.[15] All we need to say with respect to the aid record of the Arab oil countries is that if their oil exports have given them the opportunity to be of help to the community of developing countries of which they form an integral part, they have accepted the challenge and the responsibility in a commendable way. In fact, the OECD report referred to earlier considers their record as laudable. Such a testimony, coming from a major donor group — in some sense a competitor — cannot be brushed aside lightly.

Before closing the present section and chapter, it would not be out of place to put on record that the Arab oil-exporting countries have also acted with a strong sense of responsibility towards the Western industrial countries as well, by selling them more oil than they really needed to sell, by placing huge orders for goods and services with the Western countries, thus 'recycling' a large part of the oil revenues received and finally by investing and placing the largest part of their unspent revenues in Western money markets. This

*Opportunity and Responsibility*

has considerably lightened the burden of the industrial countries. Furthermore, the orders and the investments have been an activating factor for the Western economies. And finally, the responsible behaviour of the oil exporters in not moving their surpluses in speculative fashion has spared the Western currencies and money markets considerable dislocation.

In fact, if one takes all these points together, and considers at the same time the attitude of the West with regard to the basic demands of the Arabs — political and economic alike — one would be led to wonder if the Arabs have not acted with an excessive sense of international responsibility, as far as the West (particularly the United States) is concerned.[16] But, in order not to end on a negative note, one could just express the hope that, with better understanding in the West of the position and policies of the Arab oil producers, and with greater cohesiveness among the Arab countries and determination to make their rights and their interests understood and respected, there would at last come the time and situation when the sense of responsibility would be reciprocated, and genuine interdependence would be established. Only then would opportunity and responsibility, on both sides of the fence, be firmly joined and fairly balanced.

*NOTES*

## Chapter 1

1. Michael Tanzer has many points and references of direct relevance to the question of the relationship between the major oil companies and the Western governments owning them. See *The Political Economy of International Oil and the Underdeveloped Countries* (Beacon Press, Boston, 1969), particularly Chapters 3, 4, and 5 and the notes relating to them. Chapter 5 shows the symbiotic relationship between oil companies and their governments until the mid 1960s. Since then, the relationship has been less striking, not for lack of desire by the two parties, but because the international climate is not favourable to as much governmental support as before. See also David Hirst, *Oil and Public Opinion in the Middle East* (Faber and Faber, London, 1966), Chapter 1; and J.E. Hartshorn, *Oil Companies and Governments: An Account of the International Oil Industry in its Political Environment* (Faber and Faber, London, 1967) especially Chapters XIV, XVII and XXII.
2. A large number of Western politicians, oil specialists and oil company executives, and academic economists rushed to the condemnation of OPEC measures in October 1973 (and subsequently in December 1973) involving price adjustments. The grounds of condemnation were that these measures were unjustified economically and, by implication, unjustified morally as they were considered inimical to the world economy and to growth and prosperity. In addition, the opponents of these measures claimed that the handling by the oil producers of their new wealth will be irresponsible and will therefore disrupt the international monetary system. There are fewer voices shouting condemnation now in the early 1980s, but some can still be heard.
3. See *The Shorter Oxford English Dictionary* (Third Edition, Revised); also see *Cassell's English Dictionary*. The definition in the text can in essence be found in both sources.
4. Readers who wish to move nearer to technicality and detail will find Ian Seymour's *OPEC: Instrument of Change* (Macmillan, 1980)

of great value, as a thorough, well-documented, and authoritative study of OPEC's policies and activities over the 20 years 1960-80. Fadhil J. Al-Chalabi's *OPEC and the International Oil Industry: A Changing Structure* (Oxford University Press, 1980) is most valuable for the identification of oil issues and policies arising from the changing relationship between OPEC and the international oil industry. His analysis of these issues is penetrating and perceptive. Between them, these two books have spared the present author a considerable amount of effort that he would have otherwise had to put into the research of oil pricing and marketing modalities and problems.
5. Calculated from *International Financial Statistics*, April 1981 for data on total trade, and the British Petroleum Company Limited, *BP Statistical Review of the World Oil Industry 1980* for the ratios. Prices for 1979 and 1980 have been averaged.
6. See Yusif A. Sayigh, *The Arab Economy: Past Performance and Future Prospects* (Oxford University Press, 1982), especially Chapter 4, for a discussion of the place of oil and gas in the economy of the Arab region.

## Chapter 2

1. Seymour, *OPEC*, p. 216.
2. Stephen Hemsley Longrigg, *Oil in the Middle East: Its Discovery and Development* (Oxford University Press, 1968), pp. 214-15 for Saudi Arabia, and Seymour, *OPEC*, p. 216 for Kuwait.
3. Seymour, *OPEC*, p. 194, quoting the 'typical' agreement between Saudi Arabia and Standard Oil Company of California in 1933.
4. The regular reader of *Arab Oil and Gas* and *Middle East Economic Survey* (two leading oil periodicals) comes out with this clear indication.
5. Longrigg, *Oil in the Middle East*, Preface, from which the quotation comes in its full context.
6. Seymour, *OPEC*, pp. 194-5.
7. Yusif Sayigh, 'Arab Oil — A Second Look', in *The Middle East Forum* (a monthly published in Beirut), January, 1957.
8. Tanzer, *Political Economy*, has a wealth of material on the manner in which many oil deals were made, both in the text and in the notes where he relies on a wide variety of references.
9. For an appreciation of the length and the toughness of the producing countries' struggle for expensing, the reader is referred to *Arab Oil and Gas* and *The Middle East Economic Survey* (already mentioned), and to OPEC, *OPEC Official Resolutions and Press Releases 1960-1980* (Pergamon Press, 1980).

*Notes*

10. This statement is not to be understood to imply papering over the social costs and other dislocations brought about by the new era of control and increased revenues; these negative effects will be touched upon later on in the book. For specific discussion of such matters, see Arab Fund for Economic and Social Development and Organization of Arab Petroleum Exporting Countries, *Energy in the Arab World* (Proceedings of the First Arab Energy Conference March 4-8, 1979, Abu Dhabi, UAE; Kuwait, 1980), vol. 1, Robert Mabro, 'Oil Revenues and the Cost of Social and Economic Development', and Yusif A. Sayigh, 'The Social Cost of Oil Revenues', pp. 285-321 and 323-41 respectively.
11. The question of integration, both horizontal and vertical, receives clear examination and analysis in Al-Chalabi, *OPEC*.
12. However, Mexico got away with nationalisation without the penalty that was to be Iran's many years later.
13. For a detailed account of the entry of the United Sates into the oil industry in the Middle East see Longrigg, *Oil in the Middle East*. See also Michael Tanzer, *The Race for Resources: Continuing Struggles over Minerals and Fuels* (Monthly Review Press, 1980), although this book does not restrict itself to the history of the oil industry in the Middle East but has a much wider horizon.
14. There is ample evidence of this feeling of alienness as perceived by the producing countries before the 1970s. See Hirst, *Oil and Public Opinion*, and Sayigh, *The Arab Economy*. The Arab Petroleum Congresses (held under the auspices of the League of Arab States) were the forum *par excellence* for the expression of such a feeling.
15. For an account of the Suez war of 1956, see Peter Calvocoressi, *Suez Ten Years After* (Pantheon Books, New York, 1967); A. Nutting, *No End of a Lesson: The Story of Suez* (Potter, New York, 1967); T. Robertson, *Crisis: The Inside Story of the Suez Conspiracy* (Hutchinson, London, 1964); A. Beaufré, *The Suez Expedition 1956* (Praeger, New York, 1969), translated from French by R. Barry.
16. Quotation from Yusif A. Sayigh, *The Economies of the Arab World: Development since 1945* (Croom Helm, London, 1978), p. 38. The list of grievances had been collated initially by Basim Itayim, *Iraqi Oil Policy 1961-1971* (MA thesis at American University of Beirut, unpublished, 1972).
17. For development plans, see Sayigh, *The Economies*, Chapter 2, 'Iraq'. See also Abbas Alnasrawi, *Financing Economic Development in Iraq: The Role of Oil in a Middle Eastern Economy* (Praeger, New York, 1968).
18. Sayigh, *The Economies*, p. 38.
19. Sonatrach, *Sonatrach* (Algiers, 1972; in Arabic), p. 12.
20. Sayigh, *The Economies*, pp. 425-6 regarding the quotation from

Ali A. Attiga, the leading Libyan economist referred to. See Note 33 at the end of the chapter on 'Libya' for full notation.
21. Sayigh, *The Economies*, p. 438.
22. For an identification of the issue of different economic systems, see also Sayigh, *The Arab Economy*, different places in Chapters 9-11.
23. Quoted in *Middle East Economic Survey*, 22 September 1972.
24. Resolutions of the Sixteenth OPEC Conference, Vienna, 24-5 June 1968; Resolution XVI. 90, pp. 80-2 in *OPEC Official Resolutions and Press Releases*. All resolutions are from this source.
25. OAPEC, 'The Agreement of the Organisation of Arab Petroleum Exporting Countries' (the Establishing Agreement), pamphlet published in Arabic in Kuwait. See Article 2; see also Article 29 which stipulates for consultation by the Members within the Organisation for the purpose of co-ordinating positions and measures relating to the conditions and developments of the oil industry.
26. This listing is based on the contents of OAPEC's Secretary-General's Annual Reports; interviews with the senior officers of the Organisation; the relevant reports on seminars, conferences, and research publications.
27. See Marwan R. Buheiry, *U.S. Threats of Intervention Against Arab Oil: 1973-1979* (Institute for Palestine Studies Papers, no. 4, Beirut, 1980) for a well-documented and probing analysis of these threats.

## Chapter 3

1. This view is also taken by J.E. Hartshorn. See his *Objectives of the Petroleum Exporting Countries*, prepared in co-operation with *Middle East Economic Survey* and Energy Economics Research Ltd. (Middle East Petroleum and Economic Publications, Nicosia, Cyprus, 1978), pp. vi, vii, 2.
2. OAPEC, Department of Exploration and Production, 'Oil and Gas Exploration in the Arab Homeland and its Future Prospects', paper prepared for the Second Arab Energy Conference held in Doha, Qatar, in March 1982 (Arabic), p. 1.
3. Based on the first 4 papers in vol. 2 of the Proceedings of the First Arab Energy Conference, published by the Arab Fund and OAPEC (already referred to under the title *Energy in the Arab World*), pp. 7-290. The same estimate emerges from papers on consumption demand for oil submitted to the Second Arab Energy Conference of March 1982. (The papers have not been published in book form yet.) See also Petroleum Economist, *OPEC Oil Report*, Second Edition 1979, (London, 1979), pp. 6-8, where an average rate of growth of consumption in excess of 16 per cent for the late 1970s is mentioned.

## Notes

4. OAPEC, *Petroleum Exploration: Role and Prospects in Hydrocarbon Resource Development* (Kuwait, 1980), p. 9.
5. Ali Ahmed Attiga, 'General Statement for the Plenary Session of the UN Conference on New and Renewable Sources of Energy, Nairobi, Kenya, August 10-12, 1981', published as Supplement to *OAPEC Bulletin*, vol. 8, no. 2, February 1982. Dr Attiga, Secretary-General of OAPEC, had given thought to the question of bridging on earlier occasions. See, for instance, his keynote speech entitled 'Crossing the Energy Bridge' at OPEC's Seminar held in Vienna in October 1979, published in the Proceedings by OPEC under the title *OPEC and Future Energy Markets* (Macmillan, 1980), pp. 29-36.
6. For further details year by year and country by country within OAPEC, see OAPEC, *Secretary-General's Seventh Annual Report, 1980*, (Kuwait, 1980; Arabic), pp. 44-64.
7. OAPEC, *Petroleum Exploration*, pp. 16-19, especially Table 2, p. 17.
8. OAPEC, 'Oil and Gas Exploration', pp. 7-10.
9. With regard to exploration and the increase in the level of reserves, see Ali Khalifa Al Sabah (Minister of Oil of the State of Kuwait), 'Conceptual Perspective for a Long-Range Oil Production Policy', paper given at the First Oxford Energy Seminar held in September, 1979. The proceedings were published by the Seminar under the editorship of Robert Mabro under the title *World Energy Issues and Policies* (Oxford University Press, 1980). See pp. 355-8 especially. See also paper by René Ortiz, Secretary-General of OPEC at the time, entitled 'Crude Oil: Issues and Policies for Oil-Exporting Countries', especially pp. 288-91, in the same volume.
10. Petroleum Economist, *OPEC Oil Report*, p. 70-1.
11. Michael Tanzer, *The Political Economy*, pp. 26-7, 117 ff.
12. Francisco Parra, 'Exploration in the Developing Countries: Trends in the Seventies, Outlook for the Eighties', paper presented at the International Petroleum Seminar sponsored by the Institut Français du Pétrole, at Nice in March 1981.
13. For production data, see League of Arab States, Directorate-General of Economic Affairs, 'The State of Oil and Gas in the Arab Homeland', document No. 2/9/S/11 of 5 July 1980, submitted to the Joint Meeting of the Arab Ministers of Foreign Affairs and of National Economy, preparatory to the 11th Arab Summit held in Amman, Jordan, in November 1980. (Document in Arabic. See Table 1, p. 5.) Data for the 1960s are based on The British Petroleum Company Limited, *1966 Statistical Review of the World Oil Industry*, p. 19. Calculations made by the present author.
14. *BP Statistical Review of the World Oil Industry 1980*, p. 19.
15. Ali Khalifa al Sabah, 'Conceptual Perspective', *loc. cit.*
16. See, in this connection, Yusif Sayigh, 'Arab Oil Policies: Self-Interest

Versus International Responsibility', in *Journal of Palestine Studies*, vol. IV, no. 3, Spring 1975 (no. 15).
17. See OAPEC, *Secretary-General's Seventh Annual Report, 1980*, Table 35, p. 114 (based on *First Chicago Bank World Report, May-June 1980*), for estimates for 1979 and 1980. It is to be remembered that these estimates do not cover Libya and Qatar – an omission which should qualify the total upwards. The estimate for surpluses abroad as of the end of 1981 is $350 billion, as will be seen in Chapter 5 below. (The source for this last estimate is the Secretariat of the League of Arab States, obtained by the author directly.)
18. Adnan A. Al-Janabi, 'The Supply of OPEC Oil in the 1980s', in *OPEC Review*, vol. IV, no. 2, Summer 1980.
19. In this connection, see the opening address by Mana Saeed Otaiba (Minister of Oil of the United Arab Emirates) at the OPEC Seminar of October 1979, already referred to, and the keynote speech of Tayeh Abdul-Karim (Minister of Oil of Iraq) entitled 'OPEC: Challenges of the Present and Strategy for the Future', at the same Seminar, published in *OPEC and Future Energy Markets*.
20. For a broad discussion of the concept of conservation and its implications, see Al-Chalabi, *OPEC*, Part I, and his paper at the OPEC Seminar of October 1979, entitled 'The Concept of Conservation in OPEC Member Countries', and Marcello Colitti's 'Commentary'.
21. See Ali M. Jaidah's paper entitled 'OPEC Policy Options' at the same OPEC Seminar, with regard to the international dimension.
22. Resolution No. XVI. 90. See the collection of *OPEC Official Resolutions*, already referred to.
23. Quoted in *Middle East Economic Survey, MEES*, 6 June 1970.
24. *MEES*, 16 November 1981, p. 6. The *International Herald Tribune*, on 17 December 1981, p. 9, said that the official Saudi Press Agency had quoted Minister Yamani as saying that the Kingdom had 'a great potential for new oil discoveries which could double its oil reserves' from the current estimate of 173 billion barrels.
25. *OPEC Bulletin*, vol. XI, no. 2, 14 January 1980, pp. 1-11. Also Seymour, pp. 175-178; Petroleum Economist *op. cit*, p. 48; and finally Table 11 in Ch. IV below, with the sources referred to in it.
26. For a thorough discussion of the inhibiting factors, see Seymour, *OPEC*, Chapter IX, but especially p. 207. The Saudi position in this respect remains unchanged. Even as late as the spring of 1982, when OPEC agreed on production cuts to relieve the pressure of surplus supplies and depressed prices in the market, Minister Yamani was still to say that his country did not consider OPEC competent to talk about production levels. None the less he announced that the Kingdom had decided on a production cut, but with the explanation that this was to be understood as an independent

*Notes*

decision, not one in conformity with an OPEC position. (Statement made by him in Vienna on 20 March 1982, and reported in *MEES* on 29 March 1982.)

27. Seymour, *OPEC*, p. 208, and Statistical Appendix pp. 268-79.
28. See Fadhil J. Al-Chalabi on this point, 'The Concept of Conservation in OPEC Member Countries', in *OPEC Review*, vol. III, no. 3, Autumn 1979.
29. See, in this connection, the papers by Robert Mabro and Yusif Sayigh, already referred to (Note 2, Chapter 2).
30. AFESD and OAPEC, *Energy in the Arab World*, vol. 2, which is all relevant to the point in the text, except for the last two papers. See pp. 7-386.
31. For a discussion of the difficulties involved in constraining consumption and estimating the chances of conservation, see Richard Eden's paper entitled 'Energy Conservation: Opportunities, Limitations and Policies' in Robert Mabro, ed., *World Energy Issues and Policies* pp. 127-39. See also Adnan Al-Janabi, 'Oil Reserves of Exporting Countries and the Time Horizon of Their Depletion', in *Oil and Arab Cooperation* (a monthly review published by OAPEC), vol. 3, no. 3, 1977.
32. For a comprehensive examination of the complex, interrelated problems in their international context, see Peter Dorner and Mahmoud A. El-Shafie, eds., *Resources and Development: Natural Resource Policies and Economic Development in an Interdependent World* (The University of Wisconsin Press, Croom Helm, London, 1980), especially essays 1, 3, 4, and 6 in Part I of the book.
33. For such a listing, see Ali A. Attiga's paper entitled 'Global Energy Transitions and the Use of OPEC Oil', in Mahmoud Abdel-Fadil, ed., *Papers on the Economics of Oil: A Producer's View* (Oxford University Press, 1979), especially pp. 15-17.
34. This used to be a major concern even back in the 1950s and 1960s. See Hirst, *Oil and Public Opinion*, p. 11; also Tanzer, *The Political Economy of International Oil*, especially Chapters 5 and 26. Presently, pricing has come to share the area of concern with security of supplies, from the standpoint of the consumers.
35. With reference to the role of 'residual supplier', see Ali Jaidah's paper entitled 'The Pricing of Petroleum', and Adnan Al-Janabi's paper entitled 'Determinants of Long-Term Demand for OPEC Oil', both in Abdel-Fadil, ed., *Papers*, especially pp. 63, 64, and 46, 47 respectively.
36. See Robert Mabro, 'The Role of Government in Regulating Energy Markets' in *OPEC and Future Energy Markets*, with respect to government intervention. For intervention on the part of consumers, see International Energy Agency, *Energy Conservation: The Role of Demand Management in the 1980s* (OECD, Paris, 1981), p. 62.

*Notes*

37. Al-Chalabi, *OPEC*, Part II, 'The System for Marketing Crude Oil: The Fundamental Structural Transformations in World Markets of Crude Oil'.
38. *Ibid.*, p. 33.
39. *Ibid.*, pp. 50-5; Hartshorn, *Oil Companies*, pp. 32-3; Al-Janabi in Abdel-Fadil, ed., *Papers*, p. 37; and Abdul-Razzaq Mulla Hussain, 'The Role of National Companies in the Control of Oil Operations', lecture given at the 5th Programme of the Basics of the Oil and Gas Industry, conducted by OAPEC 21 February to 19 March 1981, pp. 22 and 23 and Statistical Appendix. (These sources have somewhat different estimates.) Hartshorn suggests that the majors still lifted 60-65 per cent of total OPEC exports by the late 1970s, under long-term contracts. If account is also taken of the liftings by the 'independents', then the sales by National Oil Companies, NOCs, to third parties may range only between 10 and 20 per cent of total exports. Ironically, Petromin, the Saudi NOC which is the largest institution of its kind and serves the largest oil-exporting country in the world, only marketed about 15 per cent of Saudi oil by the late 1970s. (Al-Chalabi, *OPEC*, p. 48, note 8.)
40. Al-Chalabi, *OPEC*, p. 53.
41. Mohammad Al-Imadi, 'Oil and Arab Development', in OAPEC, *Papers on Arab Oil Industry* (Kuwait, 1981; Arabic), especially pp. 289-90. What appears in the text is partly quoted, partly paraphrased.
42. Hartshorn, *Oil Companies*, p. 17.
43. See Al-Janabi in Abdel-Fadil, *Papers*. In this essay, Al-Janabi examines 14 price models and analyses several of them carefully. (See also OECD, *Economic Outlook*, December 1981, where the factor of crude prices is seen to provide insufficient explanation of changes in demand.) I have relied heavily on Al-Janabi's analysis of price and long-term demand for crude oil.
44. Al-Janabi in Abdel-Fadil, *Oil Companies*, especially pp. 41-8 which are paraphrased in a few paragraphs in the text.
45. See OPEC, *Petroleum Product Prices and their Components in Europe 1962-79*, quoted in Seymour, *OPEC*, p. 289.
46. For an examination of company profits and government taxes on oil products in consuming countries, see Louis Turner, *Oil Companies in the International System* (Royal Institute of International Affairs, London, 1978), Chapter 7.
47. Al-Janabi in Abdel-Fadil, *Oil Companies*, pp. 44-6. The quotation is from p. 48.
48. Seymour, *OPEC*, quoting *Petroleum Intelligence Weekly*, 2 September 1963 and *MEES*, 6 September 1963.
49. Quoted from Seymour, *OPEC*, pp. 285-6.
50. Ibid., p. 53.

51. *International Financial Statistics*, December 1974, shows that the index of export prices of Western Industrial Countries for 1970 was higher than that for 1969 by 5.2 per cent.
52. As selected by the author from OPEC, *OPEC Chronology 1960-1980* (Vienna, 1980), pp. 6-39. The quotations, where they occur, come from the same source. (The phrase 'turning of the tide' is Seymour's.)
53. The reference to reunification comes from 'Supplement' to *MEES*, 2 November 1981, a report by Seymour.
54. In this connection, see Seymour, *OPEC*, pp. 175-8; Chalabi, *OPEC*, pp. 93-5; and Ahmad Zaki Yamani, 'The Changing Pattern of World Oil Supplies', in Abdel-Fadil, *Papers*, especially p. 25.
55. Hartshorn, *Oil Companies*, p. 10: 'But a mere examination of the record serves to emphasize how unlike the textbook stereotype of organized cartel behaviour OPEC price determination has turned out, in practice, to be.' See also Tanzer, *The Political Economy of International Oil*, Chapter 5, Abbas Alnasrawi's paper, 'Arab Oil and the Industrial Economies: The Paradox of Oil Dependency' (*Arab Studies Quarterly*, vol., 1 No. 1) is of direct relevance to the issue under discussion.
56. This point has already been raised in the context of depletion and production policies. It would be useful to refer to some very telling statements in this connection. Al-Janabi (in Abdel-Fadil, *Papers*, p. 48) says 'If we take the criteria of foreign exchange required for fixed capital formation, we might find that present production levels are almost twice the maximum necessary.' UAE Oil Minister Otaiba said in his opening address of the OPEC Seminar of 1979 (already referred to under the title *OPEC and Future Energy Markets*, p. 4): 'The United Arab Emirates, for example, produces 1.85 million barrels per day. We do not need this amount of oil, if we have to consider only our own requirements: we should produce something in the region of 500,000 barrels a day.' Iraqi Minister of Oil Abdul-Karim expressed the same conviction in effect at the same OPEC Seminar (p. 9), by saying: 'The sacrifices of OPEC, whose burdens increase day after day, are represented by its continuing to produce oil at rates which exceed its total needs to the financial revenues derived from exports, and, consequently, converting a large part of its non-renewable oil resources to financial and monetary assets and investments, the real value of which depreciates day after day as a result of inflation and fluctuation of exchange rates, and for which no adequate guarantees are available to protect their ownership in the future.'

Finally, Saudi Minister of Oil Yamani said on BBC Television (reported in *MEES*, 22 May 1978) 'what we are doing right now is against that [our] self-interest per se ... because we are producing much more than what we need for our financial requirements, we are depleting our oil resources. Thus, we are accumulating a surplus and losing on that surplus'. On another occasion in the same month (reported in *Time* magazine on 22 May 1978) Yamani was more explicit as to the reason why the Kingdom of Saudi Arabia was producing more oil than its needs called for. Referring to the vast and expensive programme for expanding production capacity in his country, he said this programme 'is not really in our interest. It is only in the interest of the West'. Most recently, (see *MEES*, 18 January 1982) Minister Yamani has stated the Kingdom's needs as capable of being satisfied by a level of production of 6.2 million barrels per day, or 2.3 million barrels lower than the ceiling that was operative when he made his statement, but 3.3 million lower than the ceiling during 1979 and 3.8 million lower than that during 1980. (See Table 3.3.)

57. 'Report of the Group of Experts Submitted to and Approved by the Fourth Meeting of the Ministerial Committee on Long-Term Strategy' (1980; stencilled). This Report has not been published, but it has been possible to examine it.
58. Hartshorn, *Oil Companies*, p. 18.
59. Fadhil Al-Chalabi, 'Middle East Crude Oil Availabilities and World Markets' in Abdel-Fadil, *Papers*, pp. 29-34. Quotation from p. 34.
60. This is the general drift of Attiga's argument in his keynote speech at the OPEC Seminar of 1979. See *OPEC and Future Energy Markets*, pp. 30-6.
61. See *International Financial Statistics*, various issues in 1974.
62. There is general acceptance of the claim that the Arab oil countries have suffered heavy losses on their surpluses deposited or placed in Western money markets, though estimates of these losses vary. See League of Arab States, Directorate-General of Economic Affairs, *Arab Reserves Abroad*, document No. 2/7/S/11 of 5 July 1980 submitted to the Joint Meeting of the Arab Ministers of Foreign Affairs and of National Economy in preparation for the 11th Arab Summit held in Amman, Jordan, in November 1980.
63. Petroleum Economist, *OPEC Oil Report*, (p. 138), speaks of capital costs being 60 per cent higher than in the West, but the Saudi Minister of Industry and Electricity Ghazi al-Qusaibi speaks of the excess in costs as being 30 per cent. (Quoted in Hartshorn, *Oil Companies*, p. 35.)
64. *BP Statistical Review of the World Oil Industry 1980*, p. 10 for imports. Regarding excess capacity being in Western Europe but not in the United States, see Petroleum Economist, *OPEC Oil Report*, pp. 138-9.
65. OAPEC, *Secretary-General's Seventh Annual Report 1980*, Table 7, p. 70.

66. As evidenced by statements made and papers given at the Second OPEC Seminar in Vienna in October 1978.
67. Calculations made on the basis of data in *BP Statistical Review, 1980.*
68. The proceedings, including papers given, appeared in OAPEC, *The Future of the Arab Refining Industry* (Kuwait, 1976; in Arabic.)
69. See Sadegh Madjidi, 'Future Trends in Oil Refining', in *OPEC Bulletin*, vol. XI, no. 16, July 1980, for a probing discussion of these issues. Madjidi also has useful discussion of disaggregated demand for refined products, and the factors that influence demand in the United States, West Europe, and Japan.
70. Hartshorn, *Oil Companies*, pp. 33-6.
71. OAPEC, *Secretary-General's Seventh Annual Report 1980*, p. 54.
72. These were held in March 1979 and March 1982 respectively. The proceedings of the first conference have appeared in 3 volumes under the general title *Energy in the Arab World*.
73. As, for instance, OAPEC, *Selected Readings in the Oil Industry*, (Kuwait, 1979; Arabic).
74. The papers given at this symposium are collected in OAPEC, *Proceedings of the Symposium on the Ideal Utilization of Natural Gas in the Arab World, Algeria 29/6-1/7, 1980* (Kuwait, 1980; Arabic and English).
75. See OAPEC, *The Ideal Utilization*, the English part, on costing (especially pp. 43-55).
76. I must give credit for many of these points to the OAPEC staff who prepared the paper entitled 'Present Status and Future Potential of Natural Gas in the Arab World', in OAPEC, *Ideal Utilization*.
77. See *OAPEC Bulletin*, vol. VI, no. 7, July 1980, for a plea for such alignment.
78. Two very important policy-directed studies stand out in this connection: Aziz Al Watari, 'Refining and Petrochemicals: Developments in Some Oil-Exporting Countries' in Robert Mabro, ed., *World Energy Issues and Policies*; and Ali Khalifa Al-Kawari, 'The Economics of Alternative Uses of Non-Associated Natural Gas in the Arabian Gulf' in *Oil and Arab Cooperation*, vol. 6, no. 3, 1980.
79. I. Trapasso, 'Oil and Gas Downstream: Joint Cooperation and Development' (p. 3), paper submitted by ENI to the 'Colloquium on the Interdependence Model' held with OAPEC in Kuwait, 14-15 October 1981, as a sequel to the 'Seminar on Development Through Cooperation Between OAPEC, Italy, and South European Countries', held in Rome, 7-9 April 1981.
80. ENI, 'The Interdependence Model Data Bank: Industrial, Infrastructural and Agricultural Projects in Arab Petroleum Exporting Countries', Rome, October 1981, pp. 39 and 40; paper submitted to the Seminar cited in the note above. (The data cannot be con-

*Notes*

sidered exhaustive, 'but represent a highly-significant sample', as the 'Foreword' states. Furthermore, no data for Qatar were available when the paper was prepared. The total number of projects surveyed in oil and non-oil sectors was 1,472 in the six countries covered.)

81. For references to this kind of advice, see United Nations Economic Commission for Western Asia (ECWA), 'Basis for the Formulation of Strategies Pertaining to Development of the Petrochemical Industry in ECWA Region', paper submitted to Experts Group Meeting on the Petrochemical Industry in the ECWA Region, 9-12 June 1981, Vienna, Austria, sponsored jointly by ECWA and United Nations Industrial Development Organisation, UNIDO; Paper No. E/ECWA/ID/WG/5/3, May 1981), pp. 2-3. This paper is useful both for background and future policy formulation.

82. OAPEC, *Secretary-General's Seventh Annual Report 1980*, Table 20, pp. 78-9, provides detailed information for individual products. The same information, detailed according to whether the projects are complete or under construction or planning, appears in *OAPEC Bulletin*, vol. 8, no. 1, January 1982, pp. 9 and 10.

83. The Petroleum Economist, *OPEC Oil Report*, Second Edition, 1979, pp. 138-40.

84. Mahmoud Izzat in OAPEC, *The Ideal Utilization*, for the information in the text. See also p. 34 in the English part of the same volume, entitled 'The Industrial Uses of Associated Gas', Progress Report on a Study by UNIDO in cooperation with Gulf Organisation for Industrial Consulting, and OAPEC, regarding the high cost of feedstock and energy.

85. See Al-Kawari, 'The Economics', *loc. cit.*, for a very penetrating and convincing comparison and analysis of the two options: the liquefaction of gas for export versus petrochemicals. He concludes his analysis with 4 questions: a. Are the capital cost estimates realistic? b. Do the Gulf oil countries enjoy a comparative advantage with respect to the production of petrochemicals? c. Do the expectations of growth of world demand for the projected products justify the establishment of the industry in the Gulf area? d. Does the process of industrialisation in the Gulf area call for a complicated industry like petrochemicals? The analysis comes up with an affirmative answer to all 4 questions.

In this connection, see Al-Watari's paper in Mabro, *World Energy Issues*, which argues equally in favour of the establishment of petrochemical industry, in preference to the liquefaction and export of associated gas.

86. Yusif A. Sayigh, *The Arab Economy: Past Performance and Future Prospects*, Chapter 9.

87. The central message in OAPEC's studies and publications, and in

the statements of the Secretary-General and the Assistant Secretary-Generals, centres around Arab co-operation. See also Peter H. Spitz, 'Outlook for OPEC Member Countries in Downstream Operations' in *Oil and Arab Cooperation*, vol. 5, no. 2, 1979 (no. 15); the papers of Trapasso and Al-Kawari already referred to; and paper entitled 'Working Paper: Introductory Remarks and Supporting Information on Main Issues', submitted to the 'Experts Group Meeting on The Petrochemical Industry in ECWA Region 9-12 June 1981, Vienna, Austria' already referred to. (Document No. E/ECWA/ID/WG 5/4, June 1981.)

88. ECWA's, 'Working Paper', pp. 7-9, where a number of institutions and meetings are listed with emphasis placed on joint Arab action.
89. 'The Industrial Uses of Associated Gas' already referred to in Note 84 above, in *The Ideal Utilization*, Part II and Annexes. The quotation is from p. 34 in the English part of the volume.
90. See Hartshorn, *Oil Companies*, for such concerns, pp. 33-6.

## Chapter 4

1. This is true in spite of the drop in demand for OPEC oil (both because of substitution and of energy conservation) which has characterised 1981 and the first quarter of 1982 (when this was being written). This is not the place to speculate on the long-term trend of demand for oil and gas, but it seems quite warranted to expect this demand to remain strong (not much below 20 million barrels per day for OPEC as a whole) for the remainder of the century — apart from the demand for gas and irrespective of fluctuations of a short-term nature.
2. Note that only for Algeria is oil revenue a relatively small proportion of GDP (30.5 per cent); for others it ranges between 56 and 83 per cent. For this and similar types of information relating to national accounts, I have relied heavily on *The Consolidated Arab Economic Report, 1981*, prepared jointly by the Secretariat of the League of Arab States, the Arab Fund for Economic and Social Development, and the Arab Monetary Fund (1981; in Arabic). (Oil revenues as calculated by the present author for 1979 vary slightly from those reported in the *Consolidated Report*, because the author's estimates were calculated on the basis of a weighted average price of $19 per barrel for 1979. See Yusif Sayigh, 'The Integration of the Oil Sector with the Arab Economies' in *OPEC Review*, vol. IV, no. 4, Winter 1980, especially pp. 29 and 30.)
3. See Sayigh, *The Arab Economy*, Table 17, p. 87 for total exports, and for original sources.
4. Information on 1979 GDP comes from *The Consolidated Report*,

*Notes*

Table 2/7, p. 195, except for Libya, where the data relate to 1978 and have therefore been replaced by 1979 data from Arab Fund for Economic and Social Development, *National Accounts Country Tables* (Kuwait, 1980; hereafter referred to as AFESD, *Country Tables*), and except for Lebanon, where no data are available for the year 1979, and instead the present author has used data provided by Fernand Sanan, 'Liban: Les Comptes de la Nation', in *L'Economiste Arabe*, vol. XXV, no. 278, January 1982, based on a series prepared and published by the Banque du Liban (which is the central bank of Lebabon, for the years 1971-80 . The information on Italy's GDP and population is from the *International Financial Statistics*, April 1981. Population data for the Arab region are from Sayigh, *The Arab Economy*, Table 1, p. 10.

5. Many, though not all, categories of information for 1980 are not available. Consequently, it would not be very useful to include this year in the various aspects of the discussion. Furthermore, the year 1980 lies outside our focus which is the 1970s.
6. The following four paragraphs are taken or adapted from pp. 29-32 of the present author's paper in *OPEC Review*, mentioned in Note 2 above.
7. W.W. Rostow, *The Stages of Economic Growth: A Non-Communist Manifesto* (Cambridge University Press, 1966), Chapter 6.
8. The quantitative information in this and the subsequent three paragraphs is slightly different from that in the author's *OPEC Review* paper referred to above, since I rely here on *The Consolidated Report* which contains more recent data where certain adjustments have been introduced.
9. League of Arab States, Directorate-General of Economic Affairs, *Economic Conditions of the Arab Countries and Relations Among Them*, Document No. 2/S/11 of 5 July 1980, submitted to the Joint Meeting of the Arab Ministers of Foreign Affairs and of National Economy in preparation for the 11th Arab Summit Meeting of November 1980. See also the General Union of Chambers of Commerce, Industry, and Agriculture for the Arab Countries, *Arab Economic Report* (Beirut, January 1980).
10. *The Consolidated Report*, Table 2/7, p. 195 and Table 2/8 p. 196 for 1980. The ratio for 1970 is taken from *Economic Conditions of the Arab Countries* cited in Note 9 above.
11. After adjustment for Libya and Lebanon. See Note 4 above.
12. OAPEC, *Secretary-General's Seventh Annual Report 1980*, Table 29, pp. 100-1.
13. These proportions relate to 1977 or 1976. Based on data in UNCTAD, *Handbook of International Trade and Development Statistics 1979*, Part 4, Table 4.1. The estimates relevant to the discussion in the text follow:

| | |
|---|---|
| Investment in the 7 countries covered, 1979 | $62.6 billion |
| Imports of goods and services, 1979 | $80.9 |
| Imports for development 60% of total imports | $48.5 |
| Developmental imports/investment = 77% | |

14. See, for instance, Yusif A. Sayigh, 'A Critical Assessment of Arab Economic Development, 1945-1977' in UN Economic Commission for Western Asia, *Population Bulletin of the United Nations Economic Commission for Western Asia*, No. 17, December 1979. See also Georges Corm, *The Missed Development: Studies in the Arab Civilisational and Developmental Crisis* (Beirut, 1981; in Arabic). A slightly earlier critique constitutes the core of Galal A. Amin's *The Modernization of Poverty* (E.J. Brill, Leiden, 1974).
15. Reported in *MEES*, 10 March 1980.
16. In the case of Iraq, the shrinkage of outlays is further dictated by considerable shrinkage in production and export, owing to the destruction by the Iranians of a large part of oil installations during the Iraqi-Iranian war. Algeria too has been forced to make substantial cuts in earlier conceived spending programmes owing to shortage in liquidity, in addition to market conditions in general.
17. See Marwan R. Buheiry, *U.S. Threats of Intervention Against Arab Oil: 1973-1979*. Many expressions of harsh criticism of OPEC and indications of Western impatience with its policies have appeared in the Western press. For references to Western attitudes, see Hartshorn, *Oil Companies*, especially pp. 66-8 and James E. Akins, 'OPEC Actions: Consumer Reactions 1970-2000', in *OPEC and Future Energy Markets*, pp. 215-38.

    The hostility to OPEC has expressed itself over the years in a large number of articles, news reports, commentaries, and cartoons in the Western press. As recently as the last few months of 1981 and the first quarter of 1982, several articles were published revealing undisguised gloating over the troubles that OPEC was having owing to the glut in the market and the resultant depressed prices. Some spoke of the 'death' of OPEC: William Brown of the Hudson Institute, a long-standing critic of OPEC, said 'OPEC is 100 per cent dead', in *International Herald Tribune*, 18 March 1982. Robert I. Samuelson (*IHT*, on the same day) said 'Oil prices are falling and OPEC is scrambling to prevent a total collapse. If it cannot, obituary writers from New York to Tokyo will gleefully proclaim the cartel's demise.' Articles containing equally undisguised hostility can be found in *U.S. News and World Report* on 15 March 1982 and *Newsweek* 15 March 1982. (See earlier articles both in *Newsweek* and *Time* of 31 August 1981.) At the official level, a senior White House official was reported by Reuter's on 23 March 1982 as saying that after a long wait, they (the US) had begun to

*Notes*

weaken the control of OPEC. (Reported in *An-Nahar* of Beirut, 24 March 1982.) In short, neither the Western objective of weakening OPEC and bringing about its disintegration, nor the gloating over the near-achievement of this objective, is a secret. And, in addition, the climate of hostility and hatred, with concentration on the Arab oil exporters, can be specially felt and seen in the style and vocabulary used and in the obviously racist and defamatory cartoons accompanying much of the writing about Arab exporters.

18. See, for instance, article by Odeh Aburdene, Vice-President, Treasury Department, the First National Bank of Chicago, entitled 'The Financial Flows from the Oil Producing Countries of the Middle East to the US for the Period 1973-79' in *MEES*, 11 August 1980.
19. See the analysis by the Bank of England of the deployment of oil exporters' surpluses: Bank of England, *Quarterly Bulletin*, December 1981 (quoted in *MEES*, 28 December 1981).
20. According to OAPEC, *Secretary-General's Seventh Annual Report 1980*, pp. 113-14, inflation in OECD countries was 8.7 per cent and 10.8 per cent for 1979 and 1980 respectively, while the returns on Arab assets were 8.7 and 9.9 per cent for these two years. Therefore real net returns were zero for 1979 and less than 1 per cent for 1980. See also Petroleum Economist, *OPEC Oil Report*, pp. 17-18 and 137.
21. This was the title of a report in *MEES*, 27 July 1981. The institutions referred to are: Bank of England, Bank of International Settlements, Chase Manhattan Bank, First National Bank of Chicago, Morgan Guaranty Trust of New York, Occidental International Corporation, and OECD. The high and low estimates of cumulative assets come from *MEES*.
22. First National Bank of Chicago, *World Report, May-June 1980*, quoted in OAPEC, *Secretary-General's Seventh Annual Report 1980*, p. 113.
23. The writings referred to include the papers published in ECWA's *Population Bulletin* No. 17, in *OPEC Review*, and in *Energy in the Arab World*, vol. 1 – all already referred to, as well as *The Economies of the Arab World: Development Since 1945* and *The Determinants of Arab Economic Development* (both published by Croom Helm, London, 1978), and *The Arab Economy*.
24. Calculations based on Tables 2/7 and 2/8 (pp. 195 and 196 respectively) in *The Consolidated Report*, after the adjustments mentioned in Note 4 above.
25. A large number of studies have appeared since the mid-1970s on manpower movements across Arab national frontiers. The reader who is interested in the subject is referred to the extensive bibliography of Mahmoud Abdel-Fadil's *Oil and Arab Unity: The Impact of Arab Oil on the Prospects of Arab Unity and Inter-Arab Econ-*

*omic Relations* (Centre for Arab Unity Studies, Beirut, 3rd edition, 1981; in Arabic). Special mention should be made here of Nader Fergany's work on the subject, the most recent parts of which have been 'Oil and Population Change in the Arab World' in *Oil and Arab Cooperation*, vol. 7, no. 1, 1981, and 'Human Exchange among the Arab Countries', in *Al-Mustaqbal Al Arabi*, (monthly of the Centre of Arab Unity Studies, Beirut), no. 39, May 1982.

26. OAPEC, *Joint Arab Projects*, monograph prepared by Samih Mas'oud.
27. See Ali A. Attiga, 'Economic Development of Oil Producing Countries', in *OPEC Bulletin*, vol. XI, no. 11, November 1981. See also his article entitled 'Oil and Arab Development' in OAPEC, *Selected Readings in the Oil Industry* (Kuwait, 1979; in Arabic), and Mohammad Imadi's article 'Oil and Arab Development' in OAPEC, *Papers on Arab Oil Industry* (Kuwait, 1981; in Arabic).
28. In this general connection, see Sayigh, *The Economies of the Arab World*, *The Determinants of Arab Economic Development*, and *The Arab Economy*; Ali A. Attiga, 'Oil and Regional Co-operation Among the Arab Countries' in Mabro, *World Energy Issues*; and Part II, Chapters 8-10 in Dorner and Shafei, eds. *Resources and Development*. These sources have already been referred to.
29. Table 36, pp. 116-17 in OAPEC, *Secretary-General's Seventh Annual Report 1980* presents the volume of investment in projects contracted for in OAPEC countries for 1979 and 1980. In order of importance, the sectors are: transport and communications, water and electricity, housing and construction, industry and mining (including hydrocarbons and petrochemicals), education, health, and social services, and finally at the bottom, agriculture.
30. Of particular significance in this context is Antoine B. Zahlan's work, including: *Science and Science Policy in the Arab Homeland* (Centre for Arab Unity Studies, Beirut, 1979; in Arabic); list of Zahlan's writings until 1979 in the bibliography of the book just mentioned; 'Arab Technology: Manpower, Institutions, Policies', in *Al Mustaqbal Al Arabi*, March 1981, No. 25; and 'Science and Technology in the Arab World: Issues, Policies, Strategies', in *Oil and Arab Cooperation*, vol. 7, no. 4, 1981.

See also Izzeddin Salhani, 'On Technological Transformation', in *Al Mustaqbal Al Arabi*, no. 29, July 1981; and a group of three articles on 'The Arabs and Technology' in the same journal (no. 37, March 1982) by Ahmad Yusif Al-Hassan, Mohammad Al-Rashid Qureish, and Ali-id-Din Hilal.
31. See in this connection: Jeffrey B. Nugent and Pan A. Yotopoulos, 'What has Orthodox Economics Learned from Recent Experience?', in *World Development*, vol. 7, pp. 541-54; Galal A. Amin, *The Modernization of Poverty*; Galal A. Amin, 'Satisfying Basic Needs

in the Experience of Arab Development', in *Al Mustaqbal Al Arabi*, no. 5, January 1979; Mohammad Hisham Khawajkieh, 'Income Distribution and Arab (Economic) Growth', in *Al Mustaqbal Al Arabi*, no. 30, August 1981; and Sayigh, *The Arab Economy*, Part II, 'Major Issues and Tasks of Arab Development'.

32. The Arab Planning Institute, Kuwait, has undertaken extensive research on the subject of income distribution, within the constraint of the shortage of data. (However, its studies have remained stencilled monographs.) See also Amin and Khawajkieh cited in the preceding Note. It ought to be stated that, in spite of the paucity of information on income distribution in most Arab countries, careful observation, plus the few studies in hand, together suggest a highly-skewed pattern of distribution which fiscal policy has not attempted to correct in any seriousness.

33. In this connection, see Ali Tawfik Sadik, 'Impact of Oil Revenue on Fiscal and Monetary Policies in OAPEC Countries', in *Oil and Arab Cooperation*, vol. 5, no. 2, 1979. Zuhhayr Mikdashi, 'Payments Imbalances, OPEC, and Financial Markets', in *Oil and Arab Cooperation*, vol. 7, no. 3, 1981, is relevant also, but has a wider focus.

34. Calculated from AFESD, *Country Tables*, and from Table 2/8, p. 196, in *The Consolidated Report*.

35. See Wehbe Al-Bouri, 'Oil in Inter-Arab Relations', in OAPEC, *Papers in Arab Oil Industry* (Kuwait, 1981; in Arabic), for a wide-ranging view of Arab co-operation.

36. The reader is referred here to the collection of 30 papers and documents first submitted to the Joint Meeting of the Arab Ministers of Foreign Affairs and of National Economy in preparation for the 11th Arab Summit of November 1980 (all dated 5 July 1980). These papers and documents were subsequently submitted to the Summit and have appeared in a publication of limited circulation.

## Chapter 5

1. For a specific discussion of the role of NOCs in the marketing of OPEC oil, see Seymour, *OPEC*, Chapter X. For a more general treatment, see Al-Chalabi, *OPEC*, Parts II and III, Hartshorn in Mabro *World Energy Issues*. See also René G. Ortiz, 'The Future Role of National Oil Companies: in *OPEC Bulletin*, vol. XI, no. 15, June 1980; Abdel Razzak Mulla Hussain, *The Role of National Companies*; and Abdulhady Hassan Taher, 'The Future Role of the National Oil Companies' in *OPEC and Future Energy Markets*, pp. 76-87.

2. We have already mentioned the joint programmes and projects, and

institutions, in Chapter 2 above.
3. See Note 17, Chapter 4.
4. The Second Arab Energy Conference held in Qatar in March 1982 had papers projecting over 11 million barrels a day of oil consumption by the Arab countries at the turn of the century. This confirms the studies on consumption submitted three years earlier to the First Arab Energy Conference.
5. See references cited in Note 30 to Chapter 4 and, in addition, League of Arab States, Directorate-General of Economic Affairs, 'Joint Arab Policies for the Application of Technology', Document submitted to the Joint Meeting of Arab Ministers of Foreign Affairs and of National Economy.
6. For a view in depth of oil in its political context, see Tanzer, *The Political Economy of International Oil*, Chapters 5 and 24 (the latter entitled 'The Naked Politics of Oil: Oil Boycotts'); James E. Akins, 'OPEC Actions: Consumers Reactions 1970-2000', in *OPEC and Future Energy Markets*, pp. 215-38; Hartshorn, *Oil Companies*, pp. 61-74 (a chapter entitled 'OPEC: The Political Dimension', written by Ian Seymour); and, finally, Al-Bouri, 'Oil in inter-Arab Relations'.
7. On this point specifically, see Hartshorn, *Oil Companies*, especially pp. 66-8.
8. Yusif Abdallah Sayigh, *Arab Oil and the Palestine Question in the 1980s* (Institute for Palestine Studies Papers, No. 17, 1981; in Arabic).
9. Yusif Sayigh, 'Arab Oil - A Second Look' in *Middle East Forum*, January 1957.
10. Organisation for Economic Co-Operation and Development, *Development Co-operation: Efforts and Policies of the Members of the Development Assistance Committee – 1981 Review* (Paris, 1981).
11. *Ibid.*, Tables G.1 and G.2, p. 226.
12. The World Bank, *World Development Report 1981* (Washington, DC, 1981), Table 16 in Annex, p. 164.
13. *The Consolidated Report*, Tables 2/7 and 2/8, pp. 195 and 197 respectively.
14. Calculated from The World Bank, *World Development Report*, Table 1 in Annex, p. 134, for OECD countries.
15. Ibrahim F.I. Shihata, 'OPEC As A Donor Group', paper published by the OPEC Fund for International Development (Vienna, 1980). See also Ibrahim F.I. Shihata and Robert Mabro, 'The OPEC Aid Record' in *Oil and Arab Cooperation*, vol. 4, no. 1, 1978.
16. Yusif Sayigh, 'Arab Oil Policies: Self-Interest versus International Responsibility' in *Journal of Palestine Studies*, vol. IV, no. 3, Spring 1975.

# INDEX

Abu Dhabi 54, 150
Abyssinia 241
accounting 21, 23, 28-9; *see also* profits
Aden 142
Africa 4, 70; North 20n., 73
agreements: concession 9, 19-22, 25, 27-8, 31, 34-5, 46, 47, 85, participation 53, 54, 96-8, price (Tehran) 50, 112, 116, (Tripoli) 50; Evian 42; exploration (in ldcs) 77; France-Algeria oil 42-3; OAPEC 57-9
agriculture 191, 206, 210, 232
aid, foreign 15, 16, 18, 30, 76, 77, 83, 93, 102, 167, 189, 193, 198, 199, 242-6
Algeria 1n., 27, 31, 36n., 39, 40, 42-3, 51-2, 55, 60, 78, 82, 96, 101, 116, 133, 148, 150, 155, 160, 161, 182, 183, 191, 198
Anglo-Iranian Oil Co. 43
anti-trust regulations 34
Arab Petroleum Congresses 20; – – Investments Corporation 162; – – Services Co. 75-6
Arabisation 101
Aramco 27, 49, 52, 54, 122
arms imports/supplies 174, 183, 189, 198, 214, 241
Asia 4, 36n., 243
Attiga, Dr Ali A. 69-70

Bahrain 1n.
Basrah Petroleum Co. 41, 53, 54
Brazil 23n.
Britain 5, 8, 37, 38, 43, 53, 241
British Petroleum 20n., 53

Canada 5, 8, 37, 151, 155, 245
Al-Chalabi, Fadhil 12, 94
China 241
coal 66, 70, 107, 134-5, 153, 194, 239
Compagnie Française des Pétroles 20n., 53, 55
companies, oil 2, 3, 9, 11-14, 17-55, 61, 75-7, 84, 85, 87, 88, 94-100, 102-3, 106-8, 110-13, 116, 142, 149, 158, 178, 221, 223, 244; national 10, 59-60, 96-100, 220-2; 'newcomers'/independents 20, 23, 36, 44, 47, 54, 94, 96, 97
compensation 53-5
concession system *see* agreements; companies
conferences: Arab Energy 68, 150; OPEC 113-18; UN – on New and Renewable Sources of Energy 70
conservation 5, 11, 18, 30, 46, 47, 57, 63, 68, 71-2, 77-94, 101, 108, 110, 115, 133-4, 194, 239
consumption 186-8, 199; energy 71, 86, 103, 132, 137, 226, 227, domestic 68, 90-1, 137, 227
control 8, 9, 12-13, 18-61, 96, 102, 123-38, 142, 178, 198-218, 220; *see also* companies; participation
costing 23, 28-9, 40, 41, 91, 103, 131-5
Council for Arab Economic Unity 204
Cuba 241

defence spending 15, 18, 29, 204, 241; *see also* arms imports
demand, security of 239

267

## Index

Denmark 37
dependency 213-15, 231, 236; *see also* interdependence; leverage
depletion 10, 29, 33, 65, 68, 69, 71, 73, 75, 77-9, 81, 83-6, 89-91 *passim*, 96, 101, 103, 107, 108, 115, 131-3, 137-8, 220
desalination 11, 72, 149, 150, 227
devaluation 29, 117-18, 123
development 6, 7-8, 14, 15, 18, 26, 29, 32, 33, 42, 63, 69, 78, 81, 83, 89, 91, 108, 132, 137, 154, 178-218, 220, 225, 229-30, 236, 239, 240, 242–4, failings of 206-18, 230-1; Arab Fund for Economic and Social – 103; OPEC Fund for International – 242; regional 199-9, 231-3
discounts, price 28, 97-8. 100. 116
distribution 14, 28, 34, 64, 94, 99, 109, 201, 234
diversification 165, 191, 201, 205-6, 210, 217
Djibouti 36n.

Ecuador 5n.
Egypt 1n., 36n., 37, 142, 204, 240
electricity generation 11, 72, 149, 150, 227
embargoes 13, 17, 37-8, 88, 240, 241
energy, alternative 31, 66-7, 69, 83, 86, 90, 92-4, 101, 103, 107-10, 113, 134-5, 137, 226, 244
ENI 53
Europe, Western 23, 28, 142-3, 150, 163, 170; EEC 145
exchange rates 89, 116-19, 128, 136, 165, 179, 196
expatriates 60, 172-3, 187, 190-1, 203, 209
exploration 8, 10, 32, 34, 36, 37, 41, 42, 59, 63, 65-77, 83, 84, 91, 100-1, 133, 201, 206, 217, 226, 227, 238, 243-4

fees 55
fertiliser production 68, 107, 149, 150, 154, 155, 157, 160, 163-4, 168, 170-1, 226

Finland 245
First National Bank of Chicago 197
food supplies 174, 214, 241
France 37, 42-3, 155, 241

Gabon 5n.
gas 1n., 11, 14, 41, 43, 101, 139, 147-56, 158, 160, 164, 169-70, 191, 227; pricing 151-3, 155; reserves 147-8, 153-4
Gaza Strip 240
GDP 18, 182-4, 186-8, 190, 201-3, 208, 239, 245
Germany, West 37, 38
Golan Heights 240
growth, rate of 202-3, 205-6, 208, 210, 212
Gulf 51, 53, 54, 78; – Cooperation Council 160; – Organisation for Industrial Consulting 170

Hartshorn, J.E. 106, 129, 147
Holland 37, 38, 245

ideology 48-9, 127
ILO 242
Imadi, Dr Mohammad 103
imbalances, economic, social 210-12
importers, oil: developing countries 20, 61, 83, 93, 118, 126, 167, 242-3; industrial 20, 61, 92-3, 113, 135, 173-4, 224-5, 234, 238-42, 246-7; *see also individual countries*; OECD
imports, dependence on 187-9, 198, 209-10, 214
income distrubtion 208, 211
independents *see* companies
India 190
Indonesia 5n.
industrialisation 18, 30, 92, 138, 142, 154, 167, 171-5, 191, 207, 237, 242
inflation 29, 89, 103, 111, 112, 116, 119, 120, 123, 125, 128, 136, 165 178-9, 191, 196, 198, 228, 233
information, exchange of 57-8, 75, 169
infrastructure 8, 13, 32, 39, 64, 65, 157, 172-3, 202, 221

## Index

integration, national 8, 15-16, 165, 175, 199-201; oil industry, horizontal 25, 34, 94, 96, 99, vertical 24, 34, 94, 98, 99
interdependence, producer/consumer 222, 225, 228, 234, 239, 244, 246-7; regional 231-2
international economic order, reform of 167, 222, 237-8, 242; *see also* New − − −
International Monetary Fund 242
investment, abroad 15, 89, 127, 165, *see also* surpluses; domestic 90, 142, 158-60, 165, 170, 186, 189-93, 201-3; regional 205, 232-3; Arab Petroleum − Corporation 162
Iran 5n., 13, 35, 39, 40, 43, 47, 53, 78, 84, 85, 87, 95, 113, 120, 122, 124, 245
Iraq 1n., 19, 26, 27, 35, 37, 39-42, 47, 53, 54, 60, 82, 87, 88, 96, 113, 114, 150, 160, 161, 182, 183, 191, 198; − National Oil Co. 20; − Petroleum Co. 19, 20, 41-2, 47, 53, 54
Israel 16, 18, 37, 240
Italy 150, 183, 241, 245, 246

Al-Janabi, Adnan 82, 109, 110
Japan 19, 23n., 155, 163, 245
Jerusalem, East 240
Joint Arab Economic Sector (JAES) 204, 205, 216, 232
joint projects 57-9, 160, 169, 204, 205, 223
Jordan 142, 204

Al-Kawari, Ali Khalifa 170
Kuwait 1n., 19, 20, 37, 39, 53, 54, 85, 88, 101, 113, 150, 160, 161, 183, 191

Latin America 4, 70, 243
League of Arab States 204
Lebanon 142, 204
leverage 18, 23, 30, 32, 44, 127, 174, 239-42
Libya 1n., 27, 31, 36n., 40, 43-4, 51, 53-5, 78, 85, 88, 114, 115, 116, 120, 150, 160, 161, 183, 191
Longrigg, Stephen 24-5, 32

Maghreb 36
market, oil 23, 25, 31, 39, 45, 47, 54, 87, 93, 94-5, 99-100, 102-3, 193
marketing 8, 12-13, 21, 34, 35, 38-41 *passim*, 59, 78, 94-101, 157, 169, 171, 176, 217, 221
Mashreq 36
Mauritania 36n.
Mexico 5, 151, 155
Morgan Guaranty Trust 197
Morocco 36n.
Mossadegh, Dr 35
Mosul Petroleum Co. 41, 53, 54

nationalisation 12, 19, 20, 22, 38, 40-51 *passim*, 53, 54, 55, 78, 88, 95, 96, 97, 111
nationalism, Arab 199, 223
New International Economic Order (NIEO) 8, 16, 17, 63, 167, 237-8, 242
Nigeria 5n., 245
North/South dialogue 242
Norway 5, 8, 245
nuclear power 107, 134-5

OAPEC 1n., 5, 6, 14, 56, 57-9, 69, 71-3, 75, 100, 101, 141, 145, 149, 162, 168, 169, 170, 223, 235, 239; Founding Agreement 57-9; Judicial Board 58-9
OECD 66, 68, 69, 72, 104, 126, 135, 136, 137, 179, 197, 214, 234, 235, 236; and aid 244-6
*Oil and Arab Cooperation* 145, 150
Oman 1n.
OPEC 1n., 4, 5, 6, 21, 23, 39-41, 44, 46, 49-54, 56-8, 60, 83-8, 106, 140-1, 173-4, 178, 194, 223, 235, 237, 239; and aid 242-4, OFID 242; and pricing 13, 46, 82, 86-7, 102-3, 111-38, 225; and production 27, 46, 68, 83, 84, 88, 100, 114, 116, 143; 'Declaratory Statement' 50, 84-5, 115; *Oil Report* 163

## Index

Pakistan 190
Palestine 16, 18, 37, 240
Parra, Francisco 77
participation: equity 19-20, 22, 23, 38-41, 44, 45, 47-54, 57, 84-5, 88, 96-8, 115, 117; popular 6, 212-13
petrochemical industries 8, 11, 14, 32, 59, 68, 107, 139, 146, 150, 154-64, 166, 168, 170-1, 173, 191, 201, 207, 217, 221, 226, 227, 234
*Petroleum Economist* 163
Philippines 190
pipelines 13, 14, 65, 94, 139, 150
Portugal 38
premiums 131-4
prices/pricing 1, 7, 8, 10-13, 20, 21, 23, 26, 28-31, 34, 37, 39-42, 44-7, 50, 56, 57, 62-4, 70-1, 78, 79, 81-7, 91, 92, 94, 98, 100, 102-3, 111-38, 178-82, 194, 202, 217, 221, 225-8, 234, 235; agreements 50, 112, 116; 'buy-back' 52, 53, 97-8; ceilings 120, 121; indexing 111, 114, 128; two-tier 119, 125
production 1, 7, 8, 10-11, 21, 34, 35, 38-40, 43, 46, 59, 65-8, 77-94, 101, 114, 115, 197, 217, 221, 226, 227, non-OPEC 194, 235; and prices 86-7, 124-7; cutbacks in 11, 17, 37-8, 41-2, 82, 83, 88, 93, 115, 122, 240; prorationing/programming 46, 57, 84, 88, 100, 116; volume of 10-11, 18, 20, 25-7, 30, 37, 45, 47, 57, 63, 98, 121-2, 178, 180-2, 194, 202, 226, 227, 235
profits 12, 28-9, 55, 110; − sharing 23, 28, 41, 43, 47, 111, 200

Qatar 1n., 54, 158, 160, 161, 183, 191; − Petroleum Co. 54

rationing 13
refining 8, 14-15, 28, 30, 32, 34, 35, 39, 59, 92, 94-6, 99, 109, 139, 142-7, 156, 158, 164, 166, 170, 173, 191, 201, 207, 217, 221, 234
regional cooperation 8, 14, 16, 17, 63, 167-9, 175-7, 204-5, 215-17, 223, 231-3; Strategy for Joint Arab Economic Action 216-17, 232
research 58, 166, 169, 201, 232
reserves: financial *see* surpluses; oil 1, 10, 18, 55, 65-9, 71-2, 77, 80, 81, 83-6, 89, 91, 92, 101, 115, 127, 132-3, 136, 148, 226, 227, 235; R/P ratio 65-6, 71, 73, 81, 86, 89, 227
revenues 11, 20, 21, 26, 28-9, 32, 33, 35, 37, 42, 47, 62, 78, 79, 84, 85, 87, 91, 92, 101, 108, 111, 115, 127, 136, 137, 180-4, 194-8, 201, 227, 228, 237, 239, 246; recycling 92, 172, 174, 246; use of 8, 15, 17, 18, 26, 30, 63, 184-9; *see also* surpluses
royalties 29, 98, 111, 116

al Sabah, Shaikh Ali Khalifa 80, 83
Santa Fe Co. 101
Saudi Arabia 1n., 19, 27, 31, 37, 39, 47, 49, 52, 54, 85-6, 88, 113, 119-22, 124-7, 133, 150, 160, 163, 183, 191, 193
security, national 14, 33, 91, 228-9, 239; regional 15, 228-9, 231; *see also* supply
seminars, OAPEC 58, 69; Oxford 80
'Seven Sisters' 20n., 34, 39, 40, 94
Seymour, Ian 12, 19, 25-6, 52, 89, 112
Shell 20, 54
Somalia 36n.
South Africa 38
South Korea 190
Sri Lanka 190
substitution, oil 10, 100, 103, 132, 135, 194, 227, 239
supply, oil 13, 100, 234-46; disruption in 37-8, 87, 120-1; security of 92, 234-9
surpluses, revenue 29, 30, 63, 82, 89, 127, 174, 179, 195-8, 214-15, 241, 246-7
Sweden 245
Switzerland 245

*Index*

Syria 1n., 37, 142, 204, 240

take, government 21, 28-9, 31, 46, 103-4; *see also* profit-sharing; revenues
tankers 14, 32, 34, 41, 65, 94, 139, 150, 234
taxation 11, 40, 46, 47, 98, 109-10, 115, 116, 130, 211
technology 33-5, 67, 166, 172, 174, 191, 201, 212, 214, 236, 241; import of 172-3, 191, 241
terminals, loading 13, 14, 65, 94
Third World 15, 17, 22, 32, 40, 63, 76, 77, 102, 143, 144, 163, 166, 167, 171, 174, 187, 199, 201, 208, 210, 212, 220, 227, 232, 235, 237, 238, 242, 244
training 32, 33, 38-9, 57-9, 60, 150, 166, 169, 172-3, 201, 212, 232; Arab – Institute 58-9
transnational corporations 9, 157-8, 175, 176, 186, 242, 244
Tunisia 1n., 36n.

UNCTAD 242
unemployment 212
UNIDO 170, 242
United Arab Emirates 1n., 119, 124, 155, 160, 183, 191
United Nations 23, 242; Conference on Energy 70
United States 19, 23, 28, 34, 36, 37, 38, 49, 72, 73, 82, 96, 127, 143, 150, 155, 170, 198, 240, 241, 245, 246, 247

Venezuela 5n., 39, 40, 44, 88, 89, 113, 116, 122, 245

war: Iraq-Iran 87; June 1967 37, 47, 240; October 1973 38, 54, 88, 240; Suez 37
West Bank 240
World Bank 76, 242, 244

Yamani, Shaikh Ahmad Zaki 86
Yemen: North 204; South 142, 204

Zionism 37